Pūrva-pakṣa

Pūrva-pakṣa

FINE-TUNING OPPOSING VIEWS

A JOURNAL FEATURING RECENT RESEARCH
OF SCHOLARS ASSOCIATED WITH THE
BHAKTIVEDANTA INSTITUTE FOR HIGHER STUDIES

VOLUME 2
SPRING 2024

RLTA PRESS
Alachua, Florida

The publishers of this volume would like to offer their profound appreciation to the Bhaktivedanta Book Trust trustees for their dedicated support of the Richard L. Thompson Archives (RLTA) and the Bhaktivedanta Institute for Higher Studies, which has facilitated this project.

For more information about the Richard L. Thompson Archives please contact:
info@richardlthompson.com · www.richardlthompson.com

For more information about the Bhaktivedanta Institute please contact:
info@bihstudies.org · www.bihstudies.org

On the Front Cover: "Astronomy: the open air observatory at Delhi" (1797). Colored aquatint by Thomas Daniell, 1815. This is one of the five open air observatories constructed by Maharaja Sawai Jai Singh II of Jaipur in the 1720s and 1730s. Image Credit: Welcome Collection. See: https://wellcomecollection.org/works/s3dpx599.

On the Back Cover: Image of Jai Singh's Benares open air observatory used on the cover of Vedic Cosmography and Astronomy (1989) by Richard L. Thompson. Originally published in 1777 in the Philosophical Transaction of the Royal Society of London 67 (598–607), the authors reported that it may have been two centuries old. A 1793 follow-up account published in Philosophical Transactions 83 (45–49) more accurately dated that Jai Singh completed this observatory in 1739. For more information on the astronomical observatories built by Jai Singh, please see:
https://www.jantarmantar.org/

ISBN: 978-1-959829-03-4

Cover and book design:
Eight Eyes
eighteyes.com

Contents

ix Introduction

1 Modern Cosmological Evidence for an Egg-Shaped Universe
BHUPAL DEV

19 Historical Perspectives on Cosmological Complexities
JONATHAN EDELMANN

28 On the Antiquity of the Star Coordinates From Indian *Jyotiṣa Śāstra*
RICHARD L. THOMPSON

45 Time in Vedic Culture as Presented in the Teachings of A. C. Bhaktivedanta Swami Prabhupāda
CHRISTOPHER J. HAYTON

76 Phenomenology as a Cross-cultural Bridge Between Science and Religion, East and West
ALEXEY TIMOSCHUK

83 Advanced Astronomy in *Bhāgavata Purāṇa*
TIZIANO VALENTINUZZI

94 The Epistemology of Physics and Metaphysics
HOWARD RESNICK

103 The Five Elements of Matter in the *Śrīmad-Bhāgavatam* and the States of Matter in Modern Science
MICHAEL CREMO

110 ISKCON's Views on Science and Vaiṣṇavism: Comparing the 1979 *Life Comes from Life* Project with the Archival Transcripts of the Recorded Conversations
S. E. KREITZER

133 Vertical Dimension and Size of the Purāṇic Universe: Existing Interpretations and New Insights
VASYL SEMENOV AND EVGENIYA SEMENOVA

148 The Correlation Between Materials Science and the Vedic System of Sounds
FEDOR VODOLAZSKIY

152 Notes on the Contributors

The term *pūrva-pakṣa* is drawn from the Indian philosophical method of argumentation, or *vāda-śāstra,* where it denotes the available views under consideration from one's own or rival traditions. In debating manuals and the Nyāya system of philosophical reasoning, *pūrva-pakṣa* is part of a comprehensive mode of forming one's own views by tracing the grounds of one's addressee.

—Rakesh Pandey
"The *Pūrva-Pakṣa* of Modern Indian Thought" (2022)

The standard technique for refuting rival views is for a commentator to introduce the view of the opposing school, called the *pūrvapakṣa,* critique it, and then establish the perspective of his own school, the siddhānta. Naturally, the representation of the *pūrvapakṣa,* the opponent's view, was sometimes selective or partial, but the ensuing discussion will provide a flavor of the rich debate and keen dialectical interaction between schools that forced theologians to fine-tune their perspectives and kept the Indic intellectual traditions alive and fertile throughout the centuries.

—Edwin F. Bryant
The Yoga Sūtras of Patañjali (2009)

About the Bhaktivedanta Institute

Excerpt taken from literature produced for the "Life Comes From Life" conference held in Vrindavan, India, October 14–16, 1977

Bhaktivedanta Institute is a center for advanced study and research into the Vedic scientific knowledge concerning the nature of consciousness and the self. The Institute is the academic division of the International Society for Krishna Consciousness. It consists of a body of scientists and scholars who have recognized the unique value of the teachings of Krishna Consciousness brought to the West by His Divine Grace A. C. Bhaktivedanta Swami Prabhupāda. The main purpose of the Institute is to explore the implications of the Vedic knowledge as it bears on all features of human culture, and to present its findings in courses, lectures, monographs, books, and a quarterly journal.

The editors of *Pūrva-pakṣa: Fine Tuning Opposing Views, Volume 2,* would like to offer their profound appreciation to the Bhaktivedanta Book Trust trustees. Their dedicated support of the activities of the Bhaktivedanta Institute for Higher Studies facilitated the production of this volume.

Introduction

Pūrva-pakṣa: Fine-Tuning Opposing Views, is a journal featuring recent research of scholars associated with the Bhaktivedanta Institute for Higher Studies based in Gainesville, Florida. Most of the volume's offerings are edited transcripts of talks given at recent seminars and conferences hosted by the Institute. There are also a number of essays that likewise draw from various presentations.

As described in the "Introduction" to the first volume, the term *pūrva-pakṣa* refers to an ancient tradition of intellectual discourse identified with the Indian subcontinent. This tradition places emphasis on learned debate that can foster a mature understanding of not just one's own argument, but those of all participants including a perceived opponent. The "Introduction" also describes historical context for the use of the term, as well as viewpoints from contemporary scholars discussing its modern relevance. The epigraph for this volume features two such references.

The conclusion to the first volume "Introduction" notes:

> Suffice it to say, there are a number of profound aspirational objectives identified with an ancient term like *pūrva-pakṣa*. Perhaps some of them could offer "immediate practical purpose" for the present-day as well. For the purposes of this journal, the aim is to feature recent research of scholars who wish to examine the ideas and practices identified with the traditions of the Indian subcontinent, while considering their relevance in a contemporary context. ... The objective is to invite a modern *pūrva-pakṣa* involving participants confident in personal convictions, enthused to analyze from different perspectives, and open to the test of critical examination in a manner that stimulates productive scholarly discourse. Thus, an analytical discussion of the content of these articles will hopefully take wing, or *pakṣa*, and achieve a "destination" far beyond initial expectations that the readers of *Pūrva-pakṣa: Fine-Tuning Opposing Views* will find rewarding.

For more information on the activities of the Bhaktivedanta Institute, please view their website: bihstudies.org. Inquiries can be forwarded to: info@bihstudies.org.

Modern Cosmological Evidence for an Egg-Shaped Universe

Bhupal Dev
Washington University

Presented at the
Nature of Spacetime and the Evolving Universe Seminar
February 19–20, 2022

The research which this essay is based on has not been published yet and is awaiting more data for confirmation, but it is very relevant for this discourse. The basic idea is that there is growing cosmological evidence which confirms the scriptural view of an egg-shaped universe.

As a brief outline, I will first discuss the Brahmāṇḍa or Cosmic Egg concept of the universe from the scriptural perspective with specific references, and then I will present some recent cosmological evidence which supports this concept. Most of this evidence involves Cosmic Microwave Background, and I will dig deeply to dissect and analyze what the CMB data really say about the structure and origin of the universe as a whole. Namely, scientists have found some issues with the CMB experimental data conflicting with theoretical predictions, and I will show how some of those issues can be addressed by invoking this egg-shaped or ellipsoidal concept of the universe. In particular, I will describe a certain anomaly called the particle anomaly and why its existence requires a non-zero eccentricity (meaning, an ellipsoidal shape) rather than a pure spherical symmetry for the universe. Finally, I'll touch upon another very interesting issue called Hubble tension, which refers to the fact that measuring the Hubble constant by two different methods gives you two different values. This is currently problematic for cosmology, but it turns out that the concept of an egg-shaped universe also can help us resolve that Hubble tension.

I will start with Vedic and Purāṇic evidence for the concept of an egg-shaped universe. There is a famous verse from the *Ṛg Veda*, verse 10.121.1, which is as follows:

hiraṇyagarbhaḥ samavartatāgre bhūtasya jātaḥ patirekāsīta |
sa dādhāra pṛthvīṁ dhyāmutemāṃ kasmai devāyahaviṣā vidhema ||

Figure 1. *Hiraṇyagarbhaḥ*, or Cosmic Egg.

Hiraṇyagarbha was present at the beginning; He was the sole lord of created beings; He upheld this earth and heaven – let us offer worship with an oblation to the divine.

The most important word in this verse is *hiraṇyagarbhaḥ*. What does it mean? *Hiraṇya* means "gold" and *garbhaḥ* means "embryo." So, *hiraṇyagarbhaḥ* means "He who was in the golden egg as an embryo." This basically refers to Lord Viṣṇu. He is referred to as *Hiraṇyagarbhaḥ* in numerous other contexts in the scriptures, including in the *Taittirīya-saṁhitā*, verse 5.5.1.2, which invokes this particular verse in a different setting. I will now move on to Purāṇic references, including the *Śrīmad-Bhāgavatam* and the *Brahma-saṁhitā* (all emphasis in the Sanskrit is added). In the *Śrīmad-Bhāgavatam*, there are several places where this concept of the golden egg, *Hiraṇyagarbhaḥ*, has been described. I will just give two examples; the first is verse 2.5.35:

sa eva puruṣas tasmād
aṇḍaṁ nirbhidya nirgataḥ
sahasrorv-aṅghri-bāhv-akṣaḥ
sahasrānana-śīrṣavān

The Lord [Mahā-Viṣṇu], although lying in the Causal Ocean, came out of it, and dividing Himself as Hiraṇyagarbha, He entered into each universe and assumed the *virāṭ-rūpa*, with thousands of legs, arms, mouths, heads, etc.

That is the translation by Śrīla Prabhupāda: *Hiraṇyagarbhaḥ*, or "He who was in the golden egg as an embryo," entered into each universe and assumed the *virāṭ-rūpa* of thousands of legs, hands, mouths, etc. The word *aṇḍaṃ*, or "egg," is again referring to the egg-shaped concept of the universe, or "Cosmic Egg." Another well-known verse in the *Śrīmad-Bhāgavatam*, verse 3.26.52, includes the line *etad aṇḍaṃ viśeṣākhyaṁ*, translated as: "This universal egg, or the universe in the shape of an egg, is called the manifestation of material energy." This is another clear reference to the Cosmic Egg.

And then, in the *Brahma-saṁhitā*, there is this famous verse (5.35):

eko 'py asau racayituṁ jagad-aṇḍa-koṭiṁ
yac-chaktir asti jagad-aṇḍa-cayā yad-antaḥ
aṇḍāntara-stha-paramāṇu-cayāntara-stham-
govindam ādi-puruṣaṁ tam ahaṁ bhajāmi

He is an undifferentiated entity as there is no distinction between potency and the possessor thereof. In His work of creation of millions of worlds, His potency remains inseparable. All the universes exist in Him and He is present in His fullness in every one of the atoms that are scattered throughout the universe, at one and the same time. Such is the primeval Lord whom I adore.

Again, this term *jagad-aṇḍa*, which can literally be translated as "universal egg," appears twice, and then *aṇḍa* ("egg") appears again in the third line as well.

This is just a sample of the numerous scriptural references to the Cosmic Egg concept of the universe. *Aṇḍa* means "egg." *Brahmāṇḍa* means "The Cosmic Egg." So, the question is, why is the universe referred to as being in the shape of an egg in the scriptures? My point is that we can try to understand this using modern cosmology.

Now let us look at the modern cosmological view on the shape of the universe, which is the main theme of this essay. To begin with, how do we know anything about the shape of the universe? What kind of observables can we study in order to find the shape of the universe? This is a nontrivial question because we are stuck inside this universe – we cannot leave it to study it from the outside like we can with other objects. We can know the shape of a ball through holding it and seeing it from the outside. But if we were trapped inside the ball, perhaps as an ant or something similar, it would be much more difficult to understand its shape as a whole. With respect to

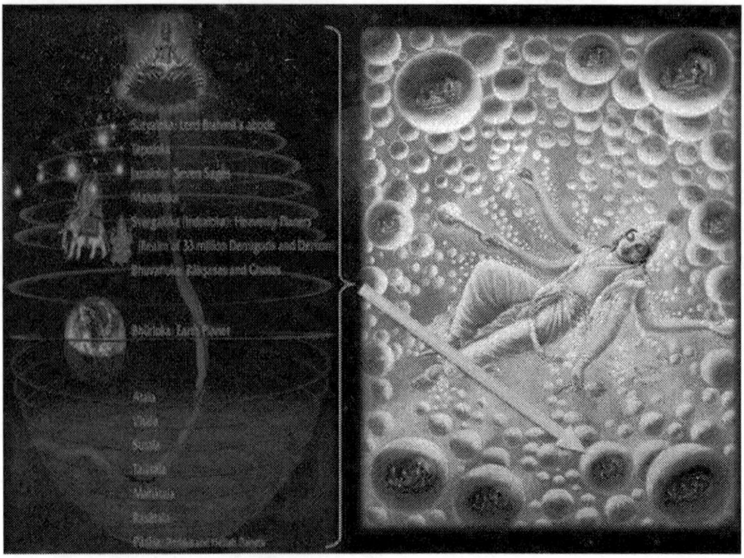

Figure 2. Right: Mahā-Viṣṇu lying in the Causal Ocean with all of the universes coming out from His pores. Left: One particular universe, which has the 14 planetary systems, the lotus stem, etc., modeled on the egg-shaped concept of the universe. [GaurangaSundar Das]

the shape of the universe, we are in the position of this ant: we cannot get outside and see what the universe looks like.

The way we try to get a holistic view of the universe is through cosmology. Modern cosmology started from theoretical mathematics. But it really came into its own once actual experimental data on Cosmic Microwave Background began coming in, which started the era of precision cosmology. Let us briefly review the history. It all began with an accidental discovery in 1964, when two scientists – a physicist named Arno Allan Penzias and a radio-astronomer named Robert Woodrow Wilson – were working at Bell Labs in New Jersey. They had a large radio antenna and they were trying to measure some radio waves, more or less for spying on the Russians since this was during the Cold War. They wanted to pick up very faint radio waves that were bouncing off of Echo balloon satellites, so they were trying to make this antenna very sensitive. They tried to remove all other possible sources of radio waves but they still always found this faint "noise" in the data. It was low but steady – background noise. They physically inspected the antenna, found some pigeon nests and droppings, removed them, and so on. Still, they could not get rid of this background noise.

Then, they talked to their colleagues in theoretical cosmology at Princeton, which was nearby. It turns out that around the same time, there was a very interesting theoretical paper being prepared by Robert Dicke, James Peebles, and others that posited the existence of detectable radiation left over from an explosion at the beginning of the universe.

They immediately realized that this could be the source of the mysterious radio noise. It was not some artificial signal – it came from the Big Bang. This was a huge discovery. They published the papers back-to-back: the theory paper first, and then the experimental paper, in the *Astrophysical Journal Letters*. The rest is history – Nobel prizes and so on. This is the origin story of precision cosmology, which really started with this discovery of the Cosmic Microwave Background. So, what exactly is CMB, and why is it so important?

To understand this, we need to think about the universe as it evolved. So, the universe is believed to have started some time ago with an explosion called the Big Bang. Now, time as we know it starts really after this very small period called "Planck time" (10^{-43} seconds) because in a period of time shorter than that, no known physics can explain what happens; and all the laws of physics break down. Before that time is called the "singularity."

So, the universal clock starts at 1×10^{-43} seconds after the Big Bang. The temperature at that time was 10^{32} kelvins, which is unimaginably hot; for reference, the surface of the sun is 5800 kelvins. We can translate this into a measurement of energy: 10^{19} GeV, or giga-electron volts. An electron volt is basically the amount of voltage you need to accelerate an electron through the potential difference of one volt. 1 eV is about 10^{-19} joules, if you want to know in terms of SI units. A giga-electron volt is 1 billion times this amount of energy, and the calculation is that the amount of energy at this time was 10^{19} GeV.

Just after the Big Bang, the whole universe was extremely tiny. It is not that the universe was in one tiny point and the rest of space was empty – all of space was basically of atomic size, with nothing existing outside of it. Then, the universe went through a phase of rapid expansion to get to a size where it could accommodate countless stars and galaxies. Why do we assume that this expansion happened rapidly? This idea, called inflation, explains why different points in the universe as we see it today are causally connected, which is indicated by the fact that the universe looks fairly isotropic and homogeneous on a pretty large scale. It turns out that if the universe had simply expanded at a constant rate from the start, it would not explain why different parts of the universe are causally connected, because light can only travel at a certain speed. If you look at the size of the universe today, it is clear that it is not possible for two points on the extreme ends to be causally connected. That is why we need this idea of inflation, which is an exponential expansion of space-time itself, and which can explain this phenomenon.

At some point after this inflation, or exponential expansion, the universe started to expand at a constant rate. After some time this rate accelerated and so on, but the point is that once inflation ended, just 10^{-35} seconds after the Big Bang, all the particles in the universe were produced. There was a theoretical particle called the "inflaton" that decayed and produced all the other particles in a process called reheating. At this point, all of these particles were in an energized cosmic soup, known as plasma. Light, which came from the Big Bang itself, could not penetrate through this primordial plasma. This is because the light interacted with all of the charged particles which were in the plasma and could not escape. The light and these particles were all trapped inside the soup together.

The visual rendition in Figure 3 shows that initially, when there were only charged particles in the primordial plasma, then photons could not escape. They just scattered against these charged particles and stayed trapped among them. The escape paths were very narrow. But, after some time, when neutral atoms formed, then most of that space became empty. The neutral particles were packed inside the atom, and the electrons were orbiting around them.

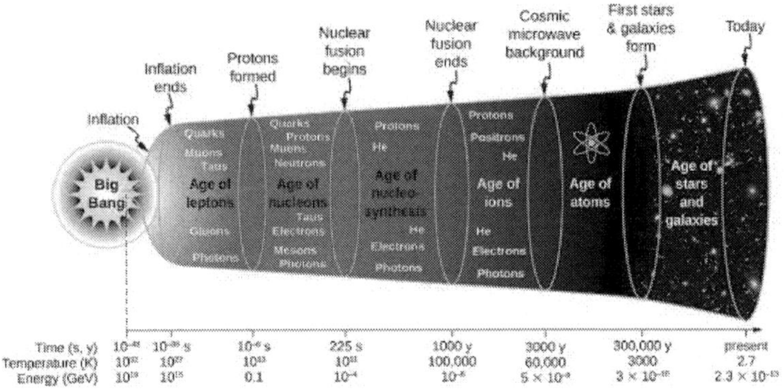

Figure 3. Era of precision cosmology.

Now, the photons could just escape through, and this is depicted with little yellow lines flying out. This is the CMB; basically, it comes from this surface, which is called the surface of last scattering, which is when the neutral atoms formed.

Then, when the universe expanded and cooled further, there was the production of neutral atoms, which are not charged. You are likely familiar with the story: an electron goes around a proton and becomes a hydrogen atom, then two of them become a helium atom, and so on, building up the periodic table. The important thing is, light does not scatter or interact as much with neutral particles as it does with charged particles. So, with the proliferation of these neutral particles, the photons could finally escape the plasma. And that is basically what we see today as the Cosmic Microwave Background. This began around 3,000 years after the Big Bang and continued until the first stars and galaxies formed around 300,000 years after the Big Bang. And so, the light we see today is a relic from that era, when the universe was just some 300,000 years old. At that time, the average temperature of the universe was still pretty high at 3000 kelvins, but then the universe expanded further and further such that today it is only 2.7 kelvins.

Now let us look at how the CMB is measured. The European Space Agency's Planck satellite, active from 2009–2013, was orbiting 1.5 million kilometers away from the earth. As the whole solar system was moving around the galaxy, it took measurements that mapped out the entire universe. It recorded data on nine different frequencies to remove what is called foreground emissions from the galactic center. This created a very precise map of the early universe. It could detect places of star formation across the galaxy by using molecular gas tracing. Its data also helped us discover many new galaxies in the universe. Its "vision" had a much deeper field of view compared to the Hubble Space Telescope because it was detecting light waves from just 380,000 years after the start of the universe all the way to today, some 13.8 billion years after the Big Bang. The Planck satellite data gives us a very nice picture of how all of these stars and galaxies and galaxy clusters formed in the early universe.

This satellite provided many beautiful pictures of what are called galaxy filaments, which are the largest structures in the universe. They are made up of superclusters – groups of galaxy clusters – and together form what is called the cosmic web. If you extend the picture backwards, then it makes a slide. On the other side of the slide, we have the CMB picture from 380,000 years after the Big Bang, and then we look forward. We can see how the light is transformed when it goes through all these filament structures, through gravitational lensing and so on. Through this analysis, we can understand more about the history of the structure formation.

This basically gave us the first-ever whole-sky map to a high degree of accuracy. That was the ultimate outcome of this data analysis and the billion-dollar project as a whole. Now that we know the context, let us try to understand exactly what this map depicts.

First of all, we know the universe is very cold – just 2.7 kelvins on average, where 0 kelvins is absolute zero. To understand why this is the case and what it has to do with the start of the universe, we need to learn about something called blackbody radiation. A black body is a perfectly opaque and nonreflective physical object, and in physics there are models for the way this type of object can emit certain kinds of electromagnetic waves depending on its temperature. If you think of a piece of black iron being heated in a fire, that

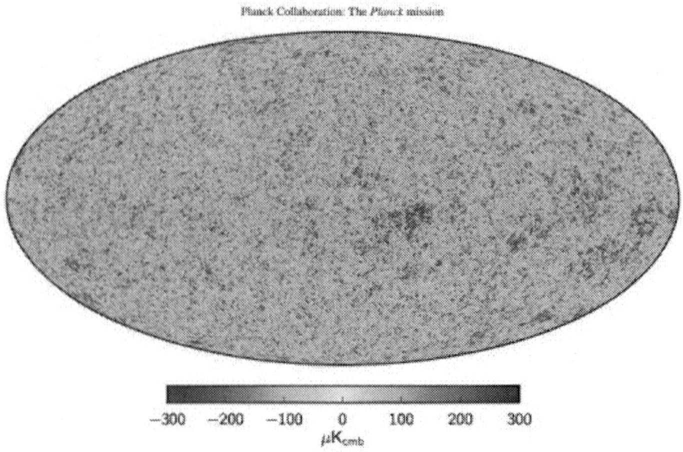

Figure 4. Planck collaboration: the *Planck* mission.

metal will absorb energy such that it becomes the same temperature as the fire, and in doing so it will start to glow some color – that is, emit electromagnetic waves – depending on how hot it is. If it is glowing red, you know that the iron is very hot, and if it is glowing white, you know it is extremely hot. This is the same principle. The range of wavelengths of this radiation is called the blackbody spectrum; some of the radiation may fall within the visible spectrum, and some of it may be invisible to our eyes. Again, this depends on the temperature of the black body.

Let us look at the relationship between temperature, radiation wavelength, and radiation intensity. For a particular temperature of the black body, we can have a graph where the x-axis is the radiation wavelength or frequency, and the y-axis is the relative intensity of the radiation. We see a peak in intensity at a particular wavelength, and then it has a characteristic spread or downward curve in the rest of the graph. This is what allows us to infer a hot piece of iron's temperature from the color of its glow – because the temperature and intensity of electromagnetic waves of various frequencies are related to each other. This relationship is expressed in Wien's displacement law, which says that the blackbody radiation curve for different temperatures will peak at different wavelengths that are inversely proportional to the temperature. So, if you decrease the temperature, then the wavelength of the light that the black body is emitting most intensely will increase proportionally, and that constant of proportionality is called the Wien constant ($b=2.897\ 768\ 551 \times 10^{-3}$ mK).

From there, we can translate blackbody temperature into its radiation frequency spectrum. In the case of 2.725 degrees Kelvin, or the average temperature of the universe, the blackbody spectrum corresponds to electromagnetic waves of around a millimeter in length, or about a hundred-megahertz frequency. This is why it is called "microwave," as megahertz is the microwave frequency range.

Now, back to our question: Why is the universe so cold today when it used to be so hot? When the CMB was being emitted, 380,000 years after the Big Bang, the universe was 3000 kelvins (2726.85 degrees Celsius). At that point, the universe was expanding, right? That means that the blackbody radiation waves were also being stretched as they traveled through space. As they were stretched, their wavelength was increasing (or their frequency was decreasing). We know from Wien's constant that there is a fixed ratio between radiation frequency and temperature, so as the wavelength increased, the temperature dropped. And that is how we know that as the universe expands, it cools down. Over 14 billion years and tremendous expansion of space-time, the average temperature of the universe went from 3000 down to 3 kelvins.

Returning to our map of the universe, the Planck satellite gave us a picture of the CMB of the whole cosmos, and we know it is very important, but what does it really tell us? First, it is a map of radiation, which we now know translates to blackbody temperature (and vice-versa). There are no street signs, but there is color coding which indicates fluctuations about the mean temperature of 2.7 kelvins. There are blue spots where the area is cooler than the mean temperature and red spots where it is warmer than the mean temperature. Overall, these cold spots and hot spots look roughly uniformly distributed. There is no major pattern to the distribution. It is pretty random, and that in and of itself is interesting. That is the signature of the CMB.

These temperature fluctuations came from quantum fluctuations in the early universe. It was actually predicted in the inflationary model that there would be small fluctuations here and there, just appearing out of the vacuum. Those fluctuations were the seeds for producing all the structure in the universe. Meaning, all of the structures that we see today, including ourselves, could not have come to exist without these fluctuations, because complete uniformity makes structure impossible. In order to have a structure like A-B-A-B, there must be A that is different from B; you cannot make A-B-A-B out of just A-A-A-A. So, these fluctuations created pockets of greater density and pockets of lesser density of matter, and then gravity started working on the pockets of greater density and brought them closer together. Eventually this formed stars, galaxies, and everything. These tiny fluctuations were very crucial for the existence of all the cosmic structure that we see today, and so it is important to measure them very precisely. From 1992 when the first measurements came in from NASA's COBE mission, to 2003 when the WMAP satellite provided much greater resolution, to the even higher quality data from the Planck satellite in 2013, the technology has evolved significantly. Maybe after

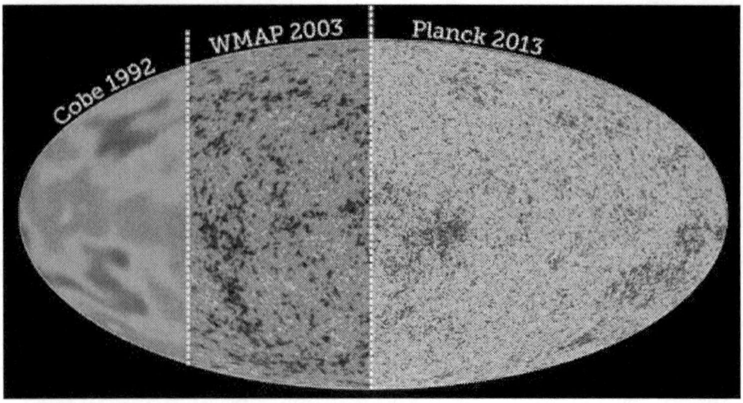

Figure 5. Increasing precision of CMB data.

10 more years, it will be even more precise. However, we are investing billions of dollars into increasing the precision of these measurements, and one might ask what we get from that investment. That is what I will discuss next.

But before I go on, just because the title of this essay is "Modern Cosmological Evidence for an Egg-Shaped Universe," let me clarify something. The cosmic maps that we see from the Planck satellite and others may look egg-shaped, but that does not mean the universe they represent is egg-shaped as well. It is impossible to tell from the map whether the universe is truly egg-shaped or not. This is simply because of the geometry of mapping the surface of a three-dimensional object onto two dimensions. It is very similar to what we do when we take a globe or 3D representation of our planet and turn it into a 2D map to hang on the wall. How do you do that?

It turns out that there are very many ways, depending on which characteristics of the 3D version are prioritized and which are less important to preserve, as a 2D map will inherently lack some fidelity to the real thing. These different kinds of maps of our planet are called projections, and the most popular ones have names like Mercator, Aitoff, Mollweide, and so forth. The Aitoff projection is an egg-shaped map of earth whose scale is most accurate along the equator and central meridian but then becomes more distorted farther away from those axes. We know that even though the Aitoff map is egg-shaped, the earth is actually a sphere. We can figure out that the earth is a sphere just from the Aitoff map, but only because we know exactly what is distorted about the projection relative to the globe, and how it is distorted.

When it comes to the cosmic map, the same basic issues come up. The satellites measure data from the whole sky which gets compiled into a 3D picture, because the universe is three-dimensional, but then it is necessary to create a 2D picture in order to present the data in an accessible and useful way. In doing so, scientists try to preserve all the features, correlations, and so on. That is why you see the cosmic map in this format, but this does not necessarily mean that the universe itself is egg-shaped, any more than the Aitoff projection means the earth is egg-shaped. But is the universe spherical or is there more to the story? To answer this question, we will dive back into temperature and mathematics.

First, we will look at multipole expansion. So, we are trying to measure the fluctuations in the temperature of the universe about some base value, because we can determine temperature from the blackbody radiation data in the cosmic map. To do this, we expand this temperature in a mathematical series of some terms that are added together; this is called the multipole expansion. These terms are usually calculated as spherical harmonics. They are functions of two angles, theta and phi, where one is the polar angle and one is the azimuthal angle, and then there are the coefficients l and m. They basically tell us which particular multipole is in what amount in this temperature fluctuation, and then we sum over all of these l's and m's, where l is similar to orbital angular momentum, and m takes integer values of $-l$ to $+l$, or $2l + 1$ values.

This is complicated, so now I will illustrate with some images. When $l = 0$, this is called a monopole. At the start of the universe, everything was all uniform with just one uniform temperature, and that is a monopole. At the next

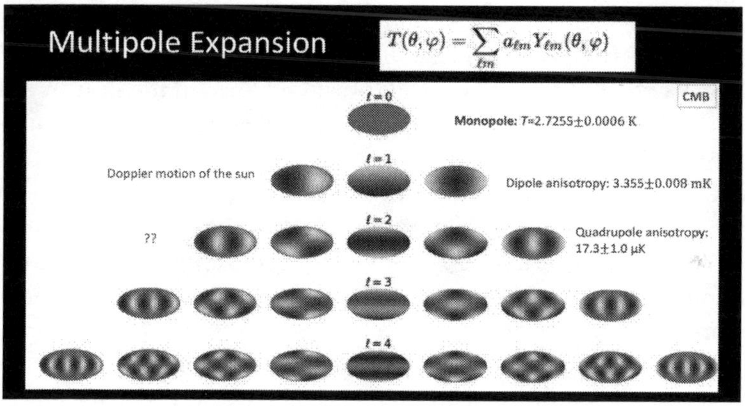

Figure 6. Multipole expansion measures the fluctuations in the temperature of the universe, developing more and more complex patterns of hot and cold.

term in the series, $l = 1$, that is called a dipole, and there are three different values because m can take a value from $-l$ to $+l$, meaning $m = -1, 0,$ or 1. But in all cases, there are two poles to the shape. With color coding, it looks like there is a half of the shape that is hot and a half that is cold, like the positive and negative poles on a magnet.

As the value of l increases, we see what are called higher poles, such as l = 2 which is called a quadrupole. In this case, there are five different possible values for m: -2, -1, 0, 1, and 2. And then if you go to l = 3, that is called an octupole. There, you have seven different values for m (namely, -3 to 3), and next is l = 4, and so on. We can calculate an infinite number of l's like this. As l keeps getting larger, the shapes develop more and more complex patterns of hot and cold. We start to see many different structures. So when we look at the cosmic map mathematically, it is actually a superposition of all these different poles. That is why this method of calculating the fluctuations is called the multipole expansion. The superposition of all of these poles together creates these structures with very small and intricate fluctuations, which look like tiny contrasting dots – cold spots and hot spots and so on.

Now, let us apply this to CMB. If l = 0, that corresponds to the monopole, which is where we say that the universe has the mean temperature 2.7 kelvins, with an error bar (or T=2.7255±0.0006 K). At l = 1 we have a dipole and there is anisotropy, meaning structural non-uniformity in different directions (like the poles of magnets). The dipole anisotropy is very small, 3.355±0.008 mK or millikelvins, which is a factor of one thousand smaller than the monopole error bar in kelvins. These are tiny fluctuations – not big. Next, the quadrupole anisotropy is even smaller: 17.3±1.0 μK or microkelvins. That is 10^{-6} K, or a million times smaller than the monopole contribution. Again, these are tiny fluctuations we are trying to measure in the temperature. But they are significant, because without them no structure could form in the universe at all.

Now this dipole term in the expansion is relatively easy to visualize in universal terms, like the monopole.

Essentially, we know that the sun is moving around the galaxy, and so, when the Planck satellite is moving around the earth and the earth is moving around the sun, it experiences what is called the Doppler effect in its light measurements. Let us quickly review the Doppler effect. If you are moving in the direction of a source of light, then the light's frequency will seem higher to you than the emitted frequency. But if you are moving away from the light, then the frequency will seem to be lower. This is experienced most commonly in sound; for example, when an ambulance passes by with its siren on, the siren sounds higher pitched as the ambulance approaches you (or you approach the ambulance) and lower pitched after it passes you and drives away. This is referred to as the change in frequency because of relative motion. So that explains the dipole anisotropy – sometimes the sun is moving towards the satellite, and sometimes it is moving away, and there is a change in the electromagnetic radiation frequency because of that.

However, there is nothing that can readily explain quadrupole anisotropy, which is where $l = 2$, or even higher modes where $l = 3$ or 4. This unsolved question is related to the question of the shape of the universe, which I will address soon. We also should keep in mind that the Planck satellite only had the resolution to see microkelvin-level anisotropies. As our technology develops and is able to detect even smaller temperature fluctuations, we may find something hiding in the higher multipoles that we do not see today.

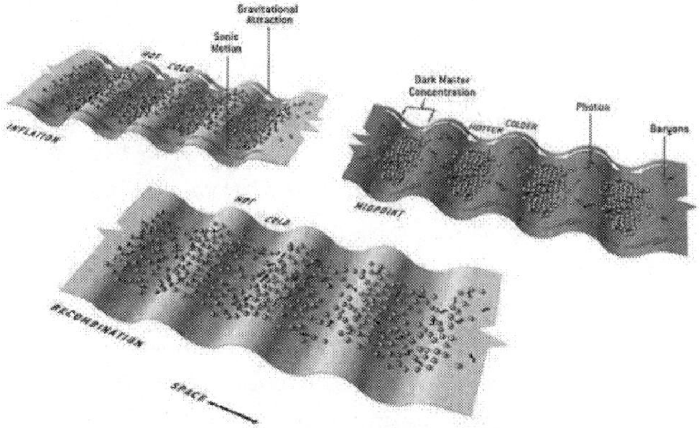

Figure 7. Plotting the theoretical predictions against the data collected by satellites matches very well over the whole spectrum, except for the lowest multipole moments.

What ultimately comes out of that cosmic map is what is called the multipole picture. We can create a graph where the x-axis is l, the multipole moment, and the Y-axis is the temperature fluctuation, or the power spectrum, because the temperature is related to how much power it has in that particular frequency band. When we plot the theoretical predictions against the data collected by satellites, they match very well over the whole spectrum, except for the lowest multipole moments.

We can also translate the x-axis scale directly into an angular scale in the sky via the very simple formula π/l. For example, for multipole moment 2 ($l = 2$), the angular scale is 90 degrees. Imagine that you are looking at two points in the sky which are 90 degrees apart – say, one at the horizon and one directly overhead. That is the largest angular scale possible. This scale depicts what are called angular correlations. For example, for $l = 10$, that will correspond to 18 degrees in the sky, and $l = 180$ will correspond to an angular scale of one degree, and so on. This π/l formula is a useful thing to remember.

There are animations of this graph which show how the power spectrum gets translated into the cosmic map we saw earlier. As we increase the multipole moment, which means going from zero to the right on the x-axis, that means we are starting with a very small multipole moment with a larger angular scale where we see bigger structures. As we go to higher and higher multipoles, we start seeing the smaller and smaller structures. Then, as we continue to a very high angular scale, we basically cannot see anything because we do not have a high enough resolution, and it all looks the same.

When we look at the graph, we see different peaks in temperature that were predicted theoretically and borne out by the data. There is a large peak starting at exactly one degree on the angular scale, and then there are three or four more peaks as the angular scale decreases, and then a gradual decline towards the x-axis. What do these peaks mean? These are called acoustic peaks. "Acoustic" means "related to sound," which could indicate a very interesting connection to the *śāstras*. In the *śāstras*, we talk about the initial or primordial sound, where all of creation started with the *oṁkāra* or *praṇava* mantra, which is *śabda-brahma*, or transcendental sound. This is something that could be of interest for future research; I am not sure if anyone has looked into it already. But let us go back to understanding what an acoustic peak is.

Imagine a simple instrument like a mouth organ or harmonica. To play this instrument, you push air into one part, then the air travels to another part, and along the way it produces harmonics. That is basically how the sound is made, and there are ways to produce nice melodies and so on, but how does that happen? When you push the air, the air gets compressed and rarefied in a wave pattern. This sound wave can either form what is called the fundamental wave, which is a half wavelength vibrating inside the column of air, or you can form overtones, which are higher frequency waves where the frequency is a multiple of the fundamental. This depends on how many compressions and rarefactions you generate in the standing wave, which you can change by stopping different holes with your fingers, blowing harder, and so on to adjust the pitch. All wind instruments generate sound in this way.

The exact same thing was happening in the early universe, and so I call it the Cosmic Symphony. We know that space and time are interrelated, so I replaced the length of this column of air in the wind instrument with time to visualize this development. Here, time starts at the early universe's inflation and goes all the way to the period called recombination, when the CMB is being emitted (380,000 years after the Big Bang). As we covered before, that is the period when neutral atoms began to form and light was able to escape the primordial plasma.

How is the early universe similar to playing a wind instrument? Let us think about what inflation does – it pushes everything outwards, very similar to someone blowing into a mouth organ. The musician produces compressions and rarefactions in the air within the instrument, and here the same thing is happening to space. Space itself is being compressed and rarefied. This produces hot and cold spots, which alternate over time like the compressed and rarefied air in a sound wave. Since the structure is the same, we can observe a fundamental wave and all the overtones or higher order harmonics, just like the wind instrument. This fundamental wave will have one hot region and one cold region, which is a dipole. Then we can start to see higher modes with patterns of multiple hot and cold spots, like the quadrupoles and octopoles we saw in the multipole expansion. As these waves all resonate across the universe like many instruments playing together, there will be some higher peaks, and so on. We basically scan over all the multipoles in this way.

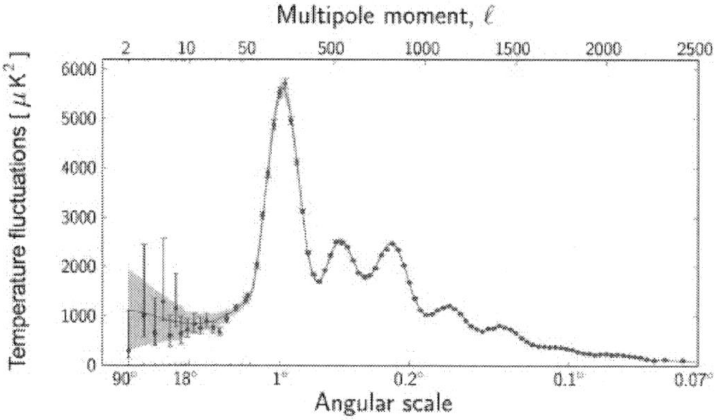

Figure 8. Cosmic Symphony: Inflation pushes everything outwards, very similar to someone blowing into a mouth organ, and the waves all resonate across the universe like many instruments playing together.

This is essentially what those peaks in the temperature power spectrum graph correspond to. The fundamental wave corresponds to the one-degree peak in temperature fluctuation, and the higher harmonics correspond to the other peaks. After some time, this oscillation dies out due to dampening, which is why we do not see more peaks. But it is still not as simple as that – there is a little more complexity because there are interactions between photons and matter. The summary, without all the mathematical details, is that there are atoms in the hot and cold regions, and matter condenses more in the cold regions because there is less kinetic energy. So, all these particles fall into something like a cold trap. If you have some dark matter in the early universe, then this trap becomes a gravitational well that particles can fall into. This process eventually produced all the structures in the universe, because this is essentially how an anisotropic structure with peaks and valleys formed at the very beginning of the universe. And then, the Cosmic Microwave Background is traveling through these valleys, which causes it to scatter, and this produces new peaks and valleys. This is the explanation for the different acoustic peaks.

Coming back to the big picture, the standard model of cosmology (Lambda-CDM) shows that the universe is made of specific proportions of certain fundamental components. Only 5% of the universe is made of ordinary matter – the stuff we can see, touch, etc. The rest is "dark," or invisible to us: 68% is dark energy, which corresponds to the Lambda, or cosmological constant, and the CDM stands for "cold dark matter," which is the other 27%. Now, if you change the proportion of these three components, then the power spectrum changes very, very dramatically, and the universe ends up very different from our own. Since we cannot measure dark matter and dark energy directly, the standard model calculates their proportions in the universe by varying these parameters and seeing what matches the experimental data.

We now have a general understanding of the standard model of Big Bang cosmology, which has been very consistent and successful so far in explaining the existence and structure of CMB (and other characteristics of the universe). However, there are still certain anomalies in CMB which do not fit into the model, and while I will not go into all the details, I will list some of them. Now, if we look at this power spectrum very carefully, we find that there are cold spots and hot spots that are distinctly separated. That is a mystery – nobody knows why this is so. Also, there seems to be a suppression of power for the large angular scales or the small multipole moments. The observed value for the quadrupole moment is 299.5 μK^2, but the expected value is 1150 ± 727 μK^2. This seems quite significant since it lies outside of the error bars. This is the kind of CMB anomaly that we cannot resolve at the moment with the standard model.

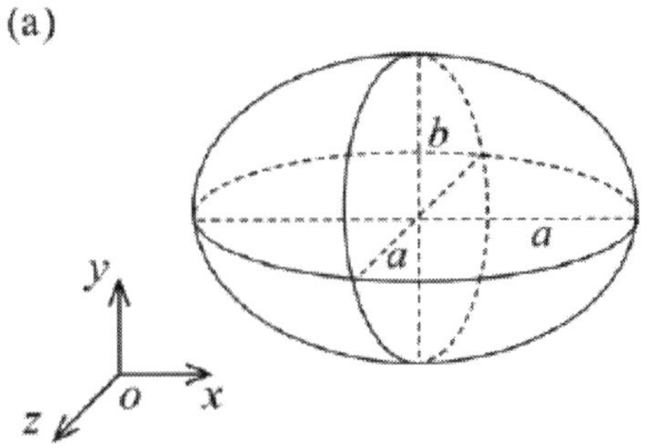

Figure 9. Ellipsoid: a shape that has at least one principal axis with a different length compared to the others.

Coming to the larger thesis of this presentation, one possible solution is through the idea of an egg-shaped, or ellipsoidal, universe. First, what is an ellipsoid? It is basically a shape that has at least one principal axis that has a different length compared to the others. If the three axes are not equal to each other, then the shape is not a sphere, because a sphere's axes are all equal. We are specifically interested in spheroid ellipsoids, where two of the axes are the same length and the third is different. We can measure the ellipticity or eccentricity of this shape by taking the ratio of two dissimilar axes, or the width and length, and then take the square root of one minus the ratio squared. This kind of shape can be called prolate or oblate, depending on whether the third axis is longer or shorter than the other two.

This measurement feeds into a special metric called the FRW (Friedmann–Robertson–Walker) metric which is based on the exact solution of the Einstein field equations of general relativity. The metric describes a homogeneous, isotropic, expanding (or otherwise, contracting) universe. If we model the shape of the universe as an ellipsoid instead of a sphere, the eccentricity will feed into this metric and modify two of the terms of the equation so that the metric can distinguish two directions of space from the third direction (just like the axes of the ellipsoid). The universe is no longer spherically symmetric. Then this modified FRW metric feeds into the quadrupole anisotropy calculation and produces a reduction of quadrupole moment to match the observed CMB best-fit value, which resolves the discrepancy between the predicted quadrupole moment and the experimental data. It turns out that if we have a small eccentricity of just 0.0067 at CMB decoupling, it does not change anything in the acoustic peaks, but at large-scale it changes things quite a bit. This is an example of how an ellipsoidal universe can explain some CMB anomalies that we cannot currently resolve with a model of the universe as spherical. The evidence for this approach and its power to resolve anomalies is growing. This seems to be consistent with the scriptural view that we have an egg-shaped universe.

Q&A

Q: With respect to the quadrupole anisotropy anomaly, are there any alternative explanations? If not, and one concludes that there is ellipticity involved, then does it mean that the photon horizon or visible horizon is ellipsoidal? If so, what are the implications for the early universe, as in quantum gravity and so on?

A: Essentially, the ellipticity concept only changes the FRW metric, and this change is very slight. Instead of an FRW metric where all three spatial directions are the same, we have in the ellipsoid case two directions that look the same and a third direction that looks a little bit different. So, remember that the ellipticity that is implied here is very small: e_{dec}=0.0067. The effect only shows up in very large scale correlations – we would not see it in a local-area effect.

Here is another way to think about it. Pragmatically speaking, we see that the earth is flat. Except over very large distances, we cannot tell that the earth's surface has a curvature. In the same way, we would not see the effect of the ellipticity of the universe except at the extreme edge of the universe. This is why we only see an effect in the very lowest multipole moments, which corresponds to the largest possible angular correlations. Aside from that, I do not think there would be any other observable effect unless we were able to look at the universe from the outside.

Q: This means the visibility of the effect is proportional to the observable volume of the universe?

A: Yes – the larger the volume we can see, the more visible the effect of ellipticity. And all of the observables we have right now do not go that far. Meaning, there is no observable that can go all the way to the edge of the universe and relay information back to us. Fundamentally, we do not have any other way of knowing whether this is true besides indirectly through the explanatory power in these modified calculations. I do not know of any explanation for the quadrupole anisotropy anomaly so far other than this particular one. It is still a mystery because nobody really knows how to explain it yet. This explanation is a promising way to resolve it, but again, we need more data to verify whether it is significant and viable.

Historical Perspectives on Cosmological Complexities

Jonathan Edelmann
University of Florida

Presented at the
Modern Journey Through Ancient Cosmography Seminar
November 21–22, 2020

In the mid-19th century, Captain John Hanning Speke (1827–1864) of the British Indian Army (re)discovered the Nile with Purāṇic maps forged by *paṇḍita* Vidyānanda working for Francis Wilford (1761–1822) in Calcutta's Asiatic Society. This was a successful practical application of information in ancient Indian texts that described Earth's geography. By extension, looking at the cosmological models described in e.g., the Fifth Canto of the *Bhāgavatam*, we may well ask, "Is the Purāṇic cosmology a map by which we can travel to Mars?" The answer is not as straightforward as a modern-day Speke might like it to be, and the reason for this has to do with understanding the purpose of such texts on their own terms and in relation to Western thought.

First, let me provide an overview of my research perspective, which is represented well in my published work. My approach in the book *Hindu Theology and Biology: The* Bhāgavata Purāṇa *and Contemporary Theory* emphasizes complexity over conflict, cohesion, and difference; science and yoga; hermeneutics, language, and meaning. I see history as opening new pathways for current and future thinking about great Hindu works and their meaning, and today these pathways for me are influenced by Indology, religious studies, philosophy, history, language, science, and the art of commentary. Western intellectual history has benefited from a rich and sophisticated conversation between theology and science, leaving us with centuries of scientific and theological literature on the subjects. Yet the Hindu traditions are virtually unused in responding to the challenging questions raised in the science and religion dialogue. The book replies to the sciences by drawing from the *Bhāgavata Purāṇa* as an important Hindu text, and in this essay I will draw from those arguments and also reference other significant Indic works and authors.

One of the greatest challenges facing Hindu traditions since the nineteenth century is their own self-understanding in light of science and technology. I hope to establish the conceptual foundations for a mutually beneficial dialogue between the Hindu theologies and the Western sciences by facing that challenge directly. Since so much of the Hinduism-science discussion is tangled in misunderstanding, it is a priority to clarify fundamental issues in each tradition, for example, the definition of consciousness, the means of generating knowledge, and the goal of knowledge itself. Objectivity is a hallmark of modern science, and I argue that this is an intellectual virtue shared by the *Bhāgavata* when it addresses topics like cosmology. Additionally, rather than taking the view that science and religion have different objects of knowledge (that is, the natural world vs. God), I argue that the *Bhāgavata* and the sciences have overlapping aims, and that many Western scientists and theologians have found science helpful in thinking about God in ways similar to that of the *Bhāgavata*.

Now, let me dive into a specific example that illustrates the way cosmology has been traditionally approached in the Hindu context. In both the *Yoga-sūtra* and the *Bhāgavata Purāṇa*, there are specific descriptions of meditative training processes that were undertaken to gain the qualifications to embark on the study of nature and the cosmos. Specifically, the *aṣṭāṅga-yoga* system with its eightfold aspects is explained in both texts. The eight aspects include restrictions, prescriptions, sitting posture, breath control, sense withdrawal, concentration, continuous concentration, and absorbed meditation. They are described in the *Yoga-sūtra* from verse 2.30 to 3.3, and they are also referenced in a similar order using very similar language in the *Bhāgavata Purāṇa* in Cantos 2, 3, and 11. The specific Sanskrit words used are the same or pretty close in both works – they are based on the same *dhātus* (roots) and *upasargas* (prefixes). The (nonexhaustive) specific chapters in the *Bhāgavatam* where you'll find these references include the first chapter of the Second

Canto, the 28th chapter of the Third Canto, and the 10th and 14th chapters in the Eleventh Canto.

According to these texts and the larger tradition, this exacting process is important for connecting with the *virāṭa-puruṣa* – the universal body of the Supreme, that is, the cosmos. And what I found is that, almost invariably, the cosmological content of the text of the *Bhāgavatam* is talked about as a form of *dhāraṇā*, which is the sixth level of the eightfold process. This level is technically the beginning stages of meditation according to the *Yoga-sūtra* – that is, after you sit properly and have your breath and your senses under control, then the cosmologies are introduced as a *dhāraṇā*, meaning "a thing that you meditate on."

Moreover, this stage is meant to prepare the mind and create a certain mental clarity and mental stability to facilitate the seventh level, *dhyāna*, or more continuous focused meditation. The Śrī Vaiṣṇava *ācārya* Ramanuja distinguished between them by saying that *dhāraṇā* is like the pouring of water, which can waver and splash, while *dhyāna* is like the pouring of oil, which flows in a steady, unbroken stream. So, *dhyāna* is a more fixed meditation on the universal body. The cosmology in these different places in the *Bhāgavatam* is meant to set the mind up for a meditation on the body of God, or the cosmos, and then progress to Viṣṇu himself. From there, more advanced meditations in *samādhi*, or fully absorbed meditation, result.

With regard to Sāṅkhya, or the metaphysical analysis of material elements, and yoga, I found a similar type of discussion and conceptualization discussed throughout the *Bhāgavatam*. One example is from 3.26.1, when Sāṅkhya is discussed by the divine incarnation Kapila. The Lord said:

> Now I shall explain to you the characteristics of the constituents of nature one by one, by knowing which the self is freed from the constraints of the qualitative forces of nature [or the *guṇas*]. I shall describe that knowledge, which they [the sages] say is the direct perception of the self and cuts knots in the heart, and which is the highest good for the self.

This is a cosmological Sāṅkhya-yoga description of the origin and the evolution of different parts of nature, but it is focused on the goal of attaining a deeper knowledge of the self and freeing oneself from the various forms of attachment. I found that this pattern is pretty well established throughout many different philosophical, proto-scientific texts in the Hindu tradition. In the *Nyāya-sūtras*, Īśvara Kṛṣṇa's *Sāṁkhya Kārikā*, Kaṇāda's *Vaiśeṣika-sūtra* – all of these texts have a similar kind of approach as the one we see here.

With regards to the Bhū-maṇḍala cosmological model in particular, I found many *Tāntric* themes associated with the subject. As a side note, the *Maṇḍala Tantra* in the classical medieval sense was simply a body of literature that had certain characteristics to it. It didn't have the exclusively and markedly sexual connotations that *Tāntra* has today. At any rate, I found much of the same language that is used in *Tāntrika* literature in the Bhū-maṇḍala discussion of the *Bhāgavatam* starting in the 16th chapter of the Fifth Canto. For example, in my book I quote Mircea Eliade discussing the *Tāntric* ideas of "entrance" and "assimilation" in connection with yogic meditation on and travel across the cosmos, and then I look at a verse from the beginning of the *Bhāgavatam* discussion of cosmology. The verse is as follows:

> Parīkṣit said: When the mind enters into the cosmic body of God, which is made of the three qualitative forces, it becomes able to enter into the most subtle form [of God] that is not [made] of the three qualitatives forces, the self-luminous Brahman, named Vāsudeva. Please describe this, O Master. (5.16.3)

The Sanskrit word that I've translated as "entering" here is *āveśita,* and this passage nicely summarizes the point that this type of meditation, or this entering into absorption in this cosmological form – this cosmology – is about preparing the mind for meditation on Viṣṇu or Brahman. It is a map of meditation. All of these cosmologies are maps of meditation in the texts. This is what I think the *Bhāgavatam* is doing; this is the *prayojana,* or the purpose, of the cosmological descriptions. This conclusion does not necessarily negate the literal, factual meaning of these Purāṇic models. It does not negate or obviate its reconstruction in a physical form, as in a planetarium, but it does orient one towards why this subject is important or how these textual items should be interpreted. This is a finding that is important to consider in the work that the Bhaktivedanta Institute is doing in this field.

And that leads to the next issue I would like to discuss here, which is hermeneutics or the interpretation of texts. Hermeneutics is a discipline within philosophy, theology, religious studies, literature – any kind of literary study – and it is given a tremendous amount of attention in the Indian tradition broadly, featuring all kinds of philosophies of language, theories of meaning, and so on. I see this as a cyclical process, which starts with what I refer to as "hearing testimony." So, we know *śravaṇaṁ kīrtanam* begins with hearing, which more broadly refers to the process of reading a text, becoming familiar with its language, studying it, understanding what the words mean and how they have been interpreted over time, and then putting that understanding into practice in some form of application. This application might be a religious practice, meditative practice, or spiritual practice – *sādhana* of any type. It also might be writing about the text or applying its insights in building or constructing something, and this practicum moves the understanding into the realm of experience. How do I make sense of this knowledge in the greater world? How am I comprehending it? How am I digesting it, so to speak? With this experience gained, the cycle begins again by going back to the text, rereading it and re-discovering it with more knowledge and more information, and then putting it into practice again. We are all going through this cyclical process continually. I see it as both a daily occurrence and also, more broadly, as a lifelong project, such that ideally one should understand texts like the *Bhāgavatam* in a different and better way – in a deeper way – after repeating this process for some time.

That is my own articulation, but I also discuss Viśvanātha's commentary of the first verse in the *Bhāgavatam* in his *Sārārtha-darśinī*, which contains similar themes. I translated a little passage from that work where he unearths a very old discussion from Ānandavardha's *Dhvanyāloka*, Abhinavagupta's *Locana*, Mammaṭa Bhaṭṭa's *Kāvyaprakāśa*, and Viśvanātha Kavirāja's *Sāhityadarpaṇa*. Viśvanātha, Jīva, Rūpa, Sanātana, Gopāla Bhaṭṭa, Baladeva Vidyābhūṣaṇa and many others seemed to be very aware of this broader secular tradition of poetics and linguistics. They wrote about these people. They applied their language and frameworks for their own work. They did not typically copy it wholesale, although Jīva actually wrote a whole book where he more or less copied the *Sāhityadarpaṇa* and then added his own commentary on it. But, largely, they use this poetic material to interpret the text of the *Bhāgavatam*, to frame it, and to make sense of it.

This is the translation of the first verse commentary in *Sārārtha-darśinī*:

> Another interpretation of *yatra tri-sargaḥ amṛṣā* is this: in whom [i.e., in the Lord, or in the *Bhāgavata*] there are three types of meaning: (1) primary, (2) indicated and (3) suggested, or [another model, which is] (1) suggestion, (2) literary excellences, and (3) linguistic ornamentation (or figures of speech) [which are used for] the creation of the literary composition. They are not false, i.e., they are real. They are wonderful because they are not of this world.

Viśvanātha Cakravartī's commentary on the first verse is very interesting because he uses these different linguistic approaches to give five different interpretations of the text (Jīva offered four different interpretations of the first verse and Viśvanātha used those and added a fifth). He engages all of these terms at various points to offer primary, indicated, and suggested meanings of the texts. He is not alone in this practice: Jīva, Rūpa, and Sanātana also took part in it. My suggestion is that there is a precedent within the Vaiṣṇava tradition – and within the Indian tradition, more broadly – for looking at a text in a variety of different ways. Not only that, these great commentators seemed to have a great deal of comfort and even a certain degree of playfulness in the way that they explored and interpreted the text. They offer, at times, several different and mutually exclusive readings of a given text.

More recently, I have discussed these kinds of approaches in an article called *Hindu Theology as Churning the Latents*. The title of this article draws largely, again, from Viśvanātha's commentary on the first verse of the *Bhāgavatam*, where he talks about textual interpretation as similar to churning the ocean of milk. This is a comparison to the story of the demigods and demons churning the ocean of milk for nectar in the Eighth Canto of the *Bhāgavatam*, and the idea he seemed to have articulated was that you keep working with the text until you draw out a certain meaning.

In my article, I argued that there are two types of theology that Viśvanātha seems to emphasize when he's doing what I call "signification," or the giving of meaning to things. When explaining this in terms of using technical knowledge of the texts to understand their meaning, I wrote, "We might call the technical aspects 'first order theology.'" Within this tradition, what we might call "first order theology" refers to learning what the text says, memorizing it, understanding the individual words, and then understanding a verse or a chapter in its entirety. This work also includes understanding how the text has been read in the past. For example, in the case of the Gauḍīya tradition, this might involve reading Śrīdhara's interpretation and reading Jīva's *Ṣaṭ-sandarbhas* and the *Bhāgavata-sandarbha*. In other words, this is the rote work of learning what the text means. In my view, it also involves learning the Sanskrit language – there is no way to get around that.

After introducing the term "first order theology," I go on to write, "It is first in the sense that the second depends on it." This means that first order theology is so named because it is a foundation for what comes next. Then: "'Second order theology' involves tasting, grasping, and internalization of the objects about which the texts speak; and it is here that the commentators discuss the overall *significance* of the theological study." This is referring to a process that I detect in the authors – the *ācāryas*, we might say, in our tradition – of learning the text and then taking that understanding and doing something with it: first saying, "This is what it says," and then moving to "This is what it means" and "This is how it fits into a bigger picture." The larger parallel that I see is that the work of the Bhaktivedanta Institute also involves both stages: first, textual reconstruction, and second, a process of signification or making sense of the text and drawing out one or more of its meanings.

The section title for this conference is "Creating a Modern Context from

Millennia-Old Discussions." I've been thinking about how to articulate the Gauḍīya Vaiṣṇava tradition in an academic world for my entire adult life, and as a conclusion I would like to present some of my current contemplations in this arena as well as past work and future directions. One important direction to explore looks at history as opening new pathways for future thinking. To illustrate this idea, let me give an example. One of my first philosophy classes as an undergraduate was a class on contemporary philosophy, starting with Immanuel Kant and working our way up to Klein and so forth. In that class, the professor did something that resonated with me then and now, namely, every time we studied the work of an 18th- or 20th-century thinker, he would go back to Plato and Aristotle and point out a question or a comment they made which addressed the same issues. His point was that there was historical precedent for all of the philosophical debates of today. People were thinking about these problems two and a half thousand years ago – and we have not stopped thinking about them.

In summary, I learned modern philosophy in a way that was inseparable from a larger historical tradition of pondering the same ideas. I was taught that the issues that people are raising today are things that people have been thinking about and writing about since we know people have been thinking and writing. This insight is very applicable to our case as well. Today's debates on cosmology, religion, presentation, interpretation, reconstruction, science, and philosophy have been dealt with before by great minds and learned people. Revisiting their thoughts on these matters is important for the general education and general knowledge of everyone discussing them today. Additionally, the understanding of similar debates in the past also creates a new language, vocabulary, and perspective for developing new projects in the future.

The second point I would like to raise in this discussion is about commentary, which I see as a distinct style of writing. Traditionally, commentary was a way for authors to connect with ancient texts and their interpretation by previous commentators over time, as well as a place to say something new. It is both minutely focused on individual words and broad in scope as a thesis is developed over and across the text. I have developed a great appreciation for the commentarial style of writing, which is represented voluminously within the Gauḍīya Vaiṣṇava tradition.

The advantage of this commentarial approach is its focus on studying the text word by word – sometimes prefix by prefix. Going back to the case of the first verse of the *Bhāgavatam*, one might spend weeks or months reading about just a handful of words. In fact, last summer I held a conference at my home in Gainesville, Florida, with a number of Vaiṣṇava scholars (Radhika Raman [Ravi M. Gupta], Krishna Kshetra Swami [Kenneth Valpey], Kiyokazu Okita, Sthita-dhi-muni [S. E. Kreitzer]) and a few of my graduate students and others. One of the conference's endeavors was exploring the commentary on the first verse. We didn't get very far in the verse – only a handful of words. It is amazing how much attention the commentators concentrated on individual words, but in doing so, they were also developing their own theoretical and philosophical conclusions, their own *vāda* and their own perspective over and across the texts. The point is, commentary provides deep insight into a text's meaning and minutiae as well as its broader sense over time.

Taking it back to the modern context, we can use this understanding of commentary from the past to think about commentary today. Much has changed in the world, and this affects our outlook and the tools we have at our disposal, but the basic questions and issues are the same. Today, commentary could be informed by Indology, religious studies, philosophy, history, studies of language and different languages themselves, science, comparative religions and theologies.

I say this because, historically speaking, each author who wrote a commentary on a text brought something of his own knowledge to it. This is a deep topic, and we do not have time to explore it extensively, but here are some brief examples: Baladeva introduced a great deal of Madhva philosophy; Viśvanātha's commentary was influenced by poetics; Jīva often cited *Advaita* and *Vedānta*; Rūpa's work also used poetics; Sanātana brought literary theory to the text. There is no compelling reason why this tradition should stop. I ask, "Why not integrate the sciences of our day into an interpretation and a commentary of the text?" And a legitimate response would be, "Well, why do it at all? What does it add to our understanding of these works?"

I have an article coming out soon where I address this question. The specific way I look at it is through the lens of: "Why would a scholar of Hinduism study European or Western thought?" In many ways, this is a response to one of my PhD mentors, Francis Clooney, who is a Jesuit priest and a great, great scholar of Vaiṣṇavism, especially South Indian Vaiṣṇavism and Tamil Vaiṣṇavism. He created and promoted a discipline called comparative theology under which he has invited Hindu scholars to study other doctrines, especially Christian thought. My article does not take a stance on this approach, but rather details a number of reasons why scholars might be for or against it. I look at this issue from the perspective of scholars of India, Hinduism, Vaiṣṇavism or Gauḍīya Vaiṣṇavism who might consider studying Western thought as they reconstruct the meaning of their own tradition. Each advantage of comparative theology is paired with a relevant disadvantage, or a reason the costs of the advantage might outweigh the benefits.

One reason a scholar of Gauḍīya Vaiṣṇavism might want to study Western thought is because it clarifies the target language, which is important in our context of introducing this philosophy to the Western world. If we are translating the *Bhāgavatam* – not only its words but also its cosmological models – it is very helpful to understand how the Western tradition will interpret it. And that is its own art and science. But one might say that a reason not to bother with clarifying the target language is that it is distracting. There simply is not enough time to fully understand what the *Bhāgavatam* is saying and, at the same time, understand the whole history of European thought.

A second reason for studying Western thought is consideration of Western hermeneutical influences. The way Westerners hear and understand the works of Gauḍīya Vaiṣṇavism may naturally be different from those raised exclusively in Hinduism because there are cultural variations in the way they learn to interpret texts. I was born and bred in New York and Boston, and I grew up reading Henry David Thoreau and listening to the Grateful Dead – all of that surely colors and influences the way I read the *Bhāgavatam* and similar works. I was also raised Protestant Christian and I have a degree in Western philosophy and studied Kant and Plato, and that surely frames how I think about the texts. There is an argument that being aware of these influences is important and valuable. But on the other side, one might say we should not invest the time and energy because there is already a need for intra-Hindu comparative religious studies. While reading Jīva, sometimes I think it is better to study Vallabha's theology, Madhva theology, Nyāya, *Raghunātha-śiromaṇi*, and Gaṅgeśa and not bother with Aquinas or Kant.

Third, one might study European thought for finding common ground and clarifying differences. One might want to use a study of Western thought to make sense of the West so they can engage with the tradition in a well-informed way. There is not so much an argument against that idea as an observation that there really is not much institutional support for this kind of work. Currently, there is no place to do it, but perhaps the Bhaktivedanta Institute could be that place.

Fourth, a Gauḍīya Vaiṣṇava scholar might find inspiration from non-Hindu traditions. Perhaps something in Western thought could inspire them. However, there is as yet no language to consider the similarities and differences between these two intellectual lineages. But if no one tries to compare them in a rigorous way, such a language will not develop.

On the Antiquity of the Star Coordinates From Indian Jyotiṣa Śāstra

Richard L. Thompson
Bhaktivedanta Institute

Presented at the
International Symposium of Ancient Indian Chronology
B. M. Birla Science Centre (Hyderabad)
January 1994

I. Introduction

A comparison is made between coordinates for 35 stars listed in traditional Indian astronomical texts (*jyotiṣa śāstra*) and the coordinates of corresponding stars listed in modern tables. I find that the error vectors pointing from the modern star positions to the corresponding *jyotiṣa* star positions are strongly correlated with the reversed proper motion vectors of the stars. Once precession is taken into account, the modern star positions show a tendency to move towards the *jyotiṣa* star positions as we go back in time.

To evaluate this, I first consider the null hypothesis, which says that we should not expect to find a significant relationship between errors in *jyotiṣa* star coordinates and proper motions of stars. I give statistical arguments showing that this hypothesis is not correct.

If there is a significant relationship between proper motions and *jyotiṣa* star coordinates, then the simplest explanation for this is that the *jyotiṣa* star coordinates were measured in the distant past. As time passed, the stars slowly moved from their positions and thereby generated error vectors pointing back along their paths. Given this hypothesis, it is possible to calculate the time of measurement of the *jyotiṣa* star coordinates. I find that these coordinates divide into a group 25,000–55,000 years old and a group less than 5,000 years old. There is also a group that cannot be clearly dated, and there is evidence suggesting that the stars in this group may not be correctly identified.

II. Jyotiṣa Star Coordinates

Several *jyotiṣa śāstra,* such as the *Sūrya-siddhānta* and the *Brahmagupta-siddhānta,* contain lists of polar ecliptic latitudes and longitudes for stars.[1] These stars include the *yogatārās* (or "principle stars") of the 28 *nakṣatras* or lunar mansions. The list in the *Sūrya-siddhānta* includes seven additional stars, making a total of 35. These are Agastya, Māgavyādha, Agni, Brahmahādaya, Prajāpati, Apāmvatsa, and Āpas.[2] I will call all these stars "*jyotiṣa* stars" and refer to them by their Sanskrit names.

Given a *jyotiṣa* star, it is natural to try to identify which star it corresponds to in the nomenclature of modern Western astronomy. I shall call this the "modern star" corresponding to the given *jyotiṣa* star. The tenth-century Muslim scholar Alberuni was perhaps the first Westerner to seek such identifications.[3] This effort was continued by scholars such as Colebrooke,[4] Burgess,[5] and Kaye[6] in the nineteenth century and B. Dikshit[7] and David Pingree[8] in the twentieth.

I have adopted the set of identifications in Table 1 for the 35 *jyotiṣa* stars. These identifications are based on the discussion given by Burgess in his translation of the *Sūrya-siddhānta.*[9] They are exactly as given by Burgess with the exception of two or three cases where he did not give a clear choice. In those cases I made an arbitrary choice from the alternatives which he offered.

We can compare a given *jyotiṣa* star position with the corresponding modern star position in the following way: First convert the polar ecliptic latitude and longitude of the *jyotiṣa* star into right ascension and declination.[10] This gives the position of the star in modern coordinates at its particular

1 Dikshit, S. B., *English Translation of Bharatiya Jyotish Sastra,* Gov. of India Press (Calcutta), 1969, pp. 338–39.

2 Burgess, Ebenezer, The Sûrya Siddhânta, Motilal Banarsidass reprint (Delhi), 1989, pp. 245, 252.

3 Sachau, E.C., trans., Alberuni's India, Kegan Paul, Trench, Trūbner & Co. (London), 1910, pp. 84– 85.

4 Colebrooke, H. T., 1807, "On the Indian and Arabian Divisions of the Zodiac," Asiatic Researches, vol. IX, pp. 323–376.

5 Burgess, pp. 202–52.

6 Kaye, G. R., *Hindu Astronomy,* Cosmo Pub. (New Delhi), 1981, p. 118.

7 Dikshit, p. 345.

8 Pingree, D. and Morrissey, P., "On the Identification of the Yogatārās of the Indian Nakṣatras," *Jour. of the History of Astronomy,* vol. 20, 1989, pp. 99–119.

9 Burgess, ibid.

10 I used the following equations to convert polar longitude and latitude (λ,β) into right ascension and declination (α,δ): $\alpha = \arctan[\cos(\varepsilon)\sin(\lambda),\cos(\lambda)]$ $\delta = \arcsin[\sin(\varepsilon)\sin(\lambda)] + \beta$

Here arctan[x,y] is the computer arctangent function, which automatically gives an angle lying in the proper quadrant. The quantity ε is the obliquity of the ecliptic. Its current value is about 23.5 degrees. However, I used 24 degrees, since this is the value used in ancient Indian astronomy, and it is the value that would have been used in India to transform star coordinates into polar longitudes and latitudes.

epoch. Then apply formulas for precession to obtain the right ascension and declination of the star in the epoch A.D. 2000.[11] These coordinates define the "*jyotiṣa* star position" in modern form. They can be compared with the right ascension and declination of the corresponding modern star listed in a catalog for the epoch A.D. 2000.[12] These latter coordinates define the "modern star position."

A great circle arc can be drawn between the star's modern position and its *jyotiṣa* position. Let D be the length of this arc in degrees. The "error vector" is a vector of length D starting at the modern position and pointing along this arc. According to Burgess, the epoch of the *jyotiṣa* star coordinates is about A.D. 490.[13] The argument for this is that if we precess the *jyotiṣa* star coordinates from various epochs to A.D. 2000, we find that the average of the lengths of the error vectors pointing from the modern star positions to the corresponding *jyotiṣa* star positions are nearly minimal for the epoch A.D. 490. Therefore I have used this epoch when converting *jyotiṣa* star coordinates to the epoch of A.D. 2000.

III. Proper Motions of Stars

Many stars are observed to move slowly with respect to the celestial sphere. These motions, called proper motions, vary from star to star. For a typical star, the proper motion amounts to a small fraction of a second of arc per year, and it is directed along a great circle arc on the celestial sphere.

In the star catalog *Sky Catalog 2000.0* (for the epoch A.D. 2000), the present rate of proper motion in both right ascension and declination is listed for each star. The negatives of these two quantities, expressed in degrees per 10,000 years, define a vector which describes the motion of the star on the celestial sphere as we follow it back in time. I will use the term "reversed proper motion vector" to refer to this vector. (For some stars the distance from the earth and the radial velocity are also listed. These quantities can be used to compute an acceleration or deceleration of the proper motion along the celestial sphere.)

I should emphasize that the right ascension and declination of a star are meaningful only in reference to a particular epoch. This is due to the gradual shifting of these coordinates caused by the precession of the equinoxes. In general, whenever I refer to a star's position, I am referring to its right

11 For calculations of precession of the equinoxes, I used the standard equations from Green, Robin M., *Spherical Astronomy*, Cambridge Univ. Press (Cambridge), 1985, p. 219 for the epoch of A.D. 2000.

12 I used Hirshfeld, A. and Sinnott, R.W., *Sky Catalogue 2000.0*, Vol. 1, Cambridge Univ. Press (Cambridge), 1982, for all modern star positions and proper motions.

13 Burgess, p. 244.

ascension and declination for the epoch A.D. 2000. This provides a standard way of specifying the position of a star on the celestial sphere. The movement of a star over thousands of years due to proper motion is also expressed in terms of right ascension and declination for A.D. 2000.

Let us suppose, for the sake of argument, that the position of a star was measured long ago with a certain error, and the results of this measurement were passed down to the present as the *jyotiṣa* coordinates for the star. As time went by, the position of the star changed as a result of the star's proper motion. Thus the error vector pointing from the modern star position to the *jyotiṣa* position must consist of two components: one due to proper motion and the other due to inevitable errors in measurement. If the error in measurement is small compared with the shift due to proper motion, then we would expect the modern star position to move towards the *jyotiṣa* position as we follow its movement backward in time.

As we track its motion backwards in time, the star follows a great circle arc that begins at the modern position and extends in the direction of the reversed proper motion vector. Another great circle arc can be drawn between the star's modern position and the *jyotiṣa* position. We shall call the angle A between these two arcs the "angle of approach." This angle lies between 0 and 180 degrees. It is the angle between the reversed proper motion vector and the error vector, as defined above.

As we follow the star's position backward in time, there comes a time when the distance between this position and the *jyotiṣa* position is minimal. Define T to be this time in years before A.D. 2000 and D_{min} to be the minimum distance in degrees. D_{min} can be compared with D, the length of the error vector pointing from the modern star position to the *jyotiṣa* star position. If the angle of approach A is greater than 90 degrees, then the star's position moves further away from the *jyotiṣa* position as we go back in time. In this case, T will be 0 and $D_{min} = D$.

The average value of D is 3.09 degrees for the 35 stars listed in the *Sūrya-siddhānta*, using the star identifications of Burgess. The *Brahmagupta-siddhānta* gives a list of coordinates for the 28 *yogatārās* of the *nakṣatras*.[14] On the whole, these coordinates seem to be more accurate than the coordinates for the 28 *nakṣatras* in the *Sūrya-siddhānta* list. We have therefore created a composite list of star coordinates by adding the seven additional stars found in the *Sūrya-siddhānta* to the *Brahmagupta-siddhānta* list. For this composite list the average value of D is 2.78 degrees. I will use this list of *jyotiṣa* star coordinates in the calculations discussed below.

Figure 1 gives an example of a modern star that moves towards its

14 Dikshit, p. 338–39

corresponding *jyotiṣa* star position as we go back in time. The modern star is Capella (Alpha Aurigae), and its present position is marked by the circle labeled "Capella." The position of Capella at successive 10,000 year intervals going into the past is marked by a series of small circles. These approach the large double circle marking the *jyotiṣa* position of Brahmahådaya at approximately 50,000 years ago (see Figure 1).

IV. Three Hypotheses

The following three hypotheses can be offered regarding possible relationships between *jyotiṣa* star coordinates and proper motions of stars:

1. Null hypothesis. There is no systematic relationship between the proper motions of stars and the error vectors pointing from modern star positions to corresponding *jyotiṣa* star positions. We therefore expect angles of approach from modern star positions to *jyotiṣa* star positions to be more or less random.

2. Historical hypothesis. The *jyotiṣa* star positions were measured in the remote past, and these stars have moved considerably since that time due to proper motions. Therefore, angles of approach tend to be smaller than they should be on the basis of chance expectation. We can learn about the time of measurement of the *jyotiṣa* star positions by computing the times T defined above.

3. Unknown alternative. There is a systematic relationship between the proper motion vectors of stars (going backwards in time) and the error vectors pointing from modern star positions to corresponding *jyotiṣa* star positions. However, the historical interpretation is incorrect, and some other explanation is required to account for this relationship.

V. The Null Hypothesis

At first glance, one would expect the null hypothesis to be correct. After all, proper motions of stars are very small, and they were measured only recently using powerful modern telescopes. They should have nothing to do with errors in naked-eye measurements made many centuries ago. However, it turns out that the null hypothesis can be ruled out by a statistical study of the data. I will begin by giving this statistical analysis.

First, one might suppose that error vectors based on *jyotiṣa* star coordinates should point in random directions relative to reversed proper motion vectors. This suggests that the angles of approach A_i (for stars i=1 to 35) should be uniformly distributed between 0 and 180 degrees. The theoretical mean for one angle selected according to the uniform probability distribution on [0,180] is $\mu = 90$ degrees, and the theoretical standard deviation is $\sigma = 180/12^{1/2} = 51.96$ degrees. The theoretical mean and variance for an average of 35 angles selected independently according to this distribution are $\mu' = 90$ and $\sigma' = \sigma/35^{1/2} = 8.78$ degrees, respectively.

If we compute the angles of approach using the star coordinates and star identifications given above, we find an average A_i of 55.78 degrees. This is about 3.9 standard deviations below $\mu' = 90$ degrees.

This finding seems to be contrary to what we might expect from the null hypothesis. However, the directions of the 35 reversed proper motion vectors are distributed nonuniformly, with a bias towards the north. Likewise, the directions of the 35 error vectors are also distributed nonuniformly. Could it be that these nonuniform distributions are responsible for the unexpectedly small angles of approach?

To answer this, it is necessary to take a deeper look at the statistics of the angles of approach. One way to do this is to assign to each star the proper motions of each of the 35 stars. This results in $35 \times 35 = 1225$ artificial "stars" which can be used to compute angles of approach. These 1225 combinations have the following properties:

(1) They have the same distributions of error vectors and reversed proper motion vectors as the 35 real stars.

(2) In the 35×35 combinations, each error vector is associated once with each proper motion vector. This erases any particular relationship between error vectors and proper motion vectors that may have existed in the original set of 35 stars.

So if the nonuniform distributions of error and reversed proper motion vectors are responsible for the small angles of approach, then we would also expect the 1225 combinations to yield small angles of approach on the average. The mean angle of approach for the 1225 artificial combinations is $\mu = 82.06$ degrees, and its standard deviation is $\sigma = 50.92$ degrees. This mean is still quite a bit higher than the average of 55.78 degrees that we found for the 35 real stars.

Given the distribution of angles created by the 1225 combinations, how probable is it that the average angle of approach will be as low as 55.78 degrees? The theoretical standard deviation for the average of 35 angles of approach chosen independently according to this distribution is $\sigma' = 50.92/35^{1/2} = 8.61$. Thus 55.78 degrees is 3.05 standard deviations below the mean of $\mu' = 82.06$ degrees. This is smaller than 3.9 standard deviations, but it is still a statistically significant deviation from the mean.

Additional evidence against the null hypothesis can be obtained by looking at certain weighted averages of the angles of approach. It turns out that if we use certain weights based on historical studies of the *nakṣatras*, we find that the statistical significance of the correlation between reversed proper motion vectors and error vectors becomes greater. I will explain this after first introducing the idea of a weighted average. A weighted average is given by the following formula:

$$\alpha = \sum_{i=1}^{N} W_i A_i \;/\; \sum_{i=1}^{N} W_i \qquad [1]$$

where the W_i's are nonnegative weights.

One set of weights measures the degree of certainty of scholars in the identification of modern stars corresponding to the *yogatārās* of the 28 *nakṣatras*. Six different scholars have given from 1 to 4 different choices for these 28 *yogatārās*.[15] The number Alt_i of choices for *yogatārā* i can be interpreted as a measure of the uncertainty of scholars regarding the identity of the *yogatārā*. Thus 4-Alt_i can be seen as a measure of the scholars' degree of certainty. The average computed using W_i = 4-Alt_i tends to emphasize those *nakṣatras* for which scholars are in agreement on the identification of the *yogatārā*.

Another set of weights represents the degree of success of Alberuni in his attempt to identify the *yogatārās*.[16] These weights, called Alb_i, are defined as follows: Alb_i = 2 if Alberuni felt he was successful in identifying the *yogatārā* i; Alb_i = 1 in the one case where he seemed to make an ambiguous statement; and Alb_i = 0 if Alberuni said he could not identify *yogatārā* i.

A third set of weights, called $Sieu_i$, has to do with correspondences between the Chinese lunar mansions called *sieus* and the *yogatārās* of the *nakṣatras*. Each Chinese *sieu* is a single star. $Sieu_i$ = 0 for *nakṣatra* i if Burgess concluded that the corresponding *sieu* does not lie in the *nakṣatra* constellation (which may consist of several stars); $Sieu_i$ = 1 if the *sieu* might be in the constellation; $Sieu_i$ = 2 if it is in the constellation but is not the *yogatārā*; and $Sieu_i$ = 3 if it is the *yogatārā*. The three sets of weights 4-Alt_i, Alb_i, and $Sieu_i$, are listed in Table 2.

Table 3 lists the results of calculating average angles of approach using these weights. In line 1 of the table, all 35 stars were used and the weights were W_i=1. In this case α is simply the average of the angles of approach for the 35 stars, and the results are as reported above.

15 Dikshit, p. 345.

16 Sachau, pp. 84–85.

In line 2, the average was restricted to the 28 *nakṣatras* and $W_i=1$. For this line, all the calculations are as before, except that 28 stars were used instead of 35. These calculations made use of the set of 28×28 = 784 artificial star combinations created by combining error vectors and proper motion vectors for the 28 *nakṣatras*. In this case the average angle of approach for the 28 *nakṣatras* turns out to be 2.17 standard deviations below the mean calculated for the 28×28 artificial combinations. (This lower value is partly due to the fact that we are using a smaller sample size.)

What happens when we look at more general weighted averages of the A_i's? If 28 A_i's are chosen independently from a distribution with mean μ and standard deviation σ, then the mean of the α given in equation (1) is simply $\mu' = \mu$. The standard deviation of this α is

$$\sigma' = \sigma \times [\sum_{i=1}^{N} W_i^2]^{1/2} \;/\; [\sum_{i=1}^{N} W_i] \qquad [2]$$

This reduces to the familiar formula $\sigma' = \sigma/N^{1/2}$ in the case where $W_i=1$. Here N=28.

Using this formula, we can compute the number of standard deviations separating the weighted average of the angles of approach and the mean calculated for the 28×28 artificial combinations. We find that for the weights W_i = 4-Alt_i and Alb_i, the statistical significance of the deviation from the mean is greater than in the case where $W_i=1$. In place of 2.17 standard deviations, we get 2.52 and 2.41, respectively.

This means that by emphasizing *nakṣatras* for which scholars were confident of their identification of the *yogatārā*, we obtain a more significant correlation between reversed proper motion vectors and vectors indicating errors in *jyotiṣa* star coordinates. Likewise, the correlation is more significant when we give emphasis to stars that Alberuni felt he could identify. A natural interpretation of this is that with these weights we are excluding (or de-emphasizing) *nakṣatras* which may have been falsely identified. By the null hypothesis, the angles of approach for falsely identified stars should not be significantly different from these for correctly identified stars (since all angles of approach should be more or less random). But this is not what we find.

Using the weights $Sieu_i$, we find that the weighted average is 2.56 standard deviations below the mean. In this case it turns out that we acquire a more significant correlation between error vectors and proper motions when we look at *nakṣatras* that are thought to correspond to Chinese *sieus*. As with 4-Alt_i and Alb_i, this set of weights may tend to select stars that are more securely identified. By the null hypothesis, these *nakṣatras* should not tend to have lower angles of approach than other *nakṣatras*, but it appears that they do.

VI. The Historical Hypothesis

Since the null hypothesis does not hold up, let us turn to the historical hypothesis. If this hypothesis is correct, then the times of closest approach for stars with small angles of approach should be estimates of how long ago the *jyotiṣa* coordinates of these stars were measured. The following string of numbers is a histogram of the times of closest approach for those stars out of the 35 that have angles of approach less than 45 degrees:

410224000110001

The first number gives the number of stars with times between 0 and 5,000 years ago. The remaining numbers give the numbers of stars with times falling in successive intervals of 10,000 years. We can see that there is one peak corresponding to the interval between 0 and 5,000 years ago and another peak corresponding to the broad interval between 25,000 to 55,000 years ago.

For the moment, let us disregard the extremely large ages represented by these peaks and try to evaluate their statistical significance. It turns out that it is easiest to do this if we restrict our attention to the 28 *nakṣatras*. For these stars, the histogram is:

410213000000000

We can create a weighted histogram using weights W_i by letting each number in the histogram be the sum of the W_i's for the stars falling in that number's time interval. If we do this for $W_i = Sieu_i$, we get:

000536000000000

It turns out that the five stars making up the more recent peak all have $Sieu_i = 0$ (meaning that none are Chinese *sieus*), and thus this peak does not appear in the weighted histogram. These stars are Punarvasu (Beta Geminorum), Maghā (Alpha Leonis), Svātī (Alpha Bootis), Abhijit (Alpha Lyrae), and Śravaṇa (Alpha Aquilae). They are all prominent stars with high proper motions, and they all fall in the set of nine *nakṣatras* with magnitudes less than 1.5. Note that the probability that five randomly chosen *nakṣatras* will fall in this group of nine is less than $(9/28)^5$. (In fact, it is 1/780.)

Given the null hypothesis, this histogram peak made of prominent stars must simply be a product of chance. However, the historical hypothesis provides a simple explanation for it. We can argue that since these stars are fast moving and prominent to the eye, their positions must have been measured relatively recently. Otherwise, they would have moved so far from their recorded positions that those positions would have been obviously in error. Burgess argues that the observed correspondences between *nakṣatras* and *sieus* must be due to cultural contact between Indian and Chinese astronomers. We can hypothesize that the stars in the recent peak do not correspond to *sieus* because they were added to the *nakṣatra* system after this period of close cultural contact. Although this explanation is certainly speculative, it does provide possible reasons for a pattern which, according to the null hypothesis, must be entirely due to chance.

Now let us consider the histogram peak in the interval from 25,000–55,000 years ago. This peak is a concentration of ages around a central point of 40,000 years. The question is: Is this concentration of ages statistically significant, or is it just a product of chance?

We can answer this question by taking advantage of the fact that $Sieu_i$ cancels the recent peak in the histogram and leaves the 25,000–55,000 year peak. This gives us a criterion for singling out the older peak which is independent of our findings regarding proper motions and error vectors. (It is independent because $Sieu_i$ depends only on Burgess's study.) Let $D4_i$ be the distance in degrees between the ith *jyotiṣa* star position and the position occupied by the corresponding modern star 40,000 years ago. If the 25,000–55,000 year peak is significant, then the weighted average of the $D4_i$'s using $W_i = Sieu_i$ should be unexpectedly small.

This turns out to be true. The weighted average of the $D4_i$'s using $Sieu_i$ as weights is 1.47 standard deviations below the mean for this average calculated using the 784 artificial stars. If we use the weights $Sieu_i \times Alb_i$ which reflect both the correspondence with Chinese *sieus* and Alberuni's star identifications, the corresponding figure becomes 1.7 standard deviations. Thus we can conclude that the 25,000–55,000 year peak has some statistical significance, but it is not highly significant.

However, there is additional evidence suggesting that this peak should be taken seriously. If we observe the individual diagrams showing the relation between modern star positions and *jyotiṣa* position, we can find several instances in which the *yogatārā* identified by Burgess does not closely approach the *jyotiṣa* position as we go back in time over an 80,000 year period. However, in some of these instances there is another star in the *nakṣatra* constellation that does closely approach the *jyotiṣa* position in this time period. Table 4 lists these new star identifications for Bharaṇī, Māgaśīrṣa, Puṣya, Pūrvaphalgunī, and Mūla.

If we adopt these five new *yogatārā* identifications, what effect does this have on the histogram of ages? Does it tend to disperse the two age peaks that we have already noticed, or does it accentuate them? The age histograms for all 35 stars and for the 28 *nakṣatras* become:

410226110110001

410215110000000

If we compare these with the corresponding histograms for the pure Burgess star identifications, we find that the two age peaks remain, and that the number of stars in the 45,000 –55,000 year range has become greater. With these modified star identifications, we can again calculate the weighted average of the $D4_i$'s using the $Sieu_i$'s as weights. This weighted average is now 2.0 standard deviations below the mean based on the 784 artificial stars. If we use the $Sieu_i \times Alb_i$'s as weights, this figure becomes 2.3 standard deviations.

These results indicate that the 25,000 –55,000 year peak has become more prominent as a result of the new star identifications. These identifications were made on the basis that the chosen star should closely approach the *jyotiṣa* star position as one goes back in time. They were not specifically chosen so that it would approach the *jyotiṣa* position in the interval of 25,000–55,000 years ago. Yet the ages of closest approach for the five new identifications are 40.7, 53.5, 73.9, 64.4, and 48.2 thousand years, respectively. Thus they do fall on the high side of this peak.

Figure 2 illustrates the choice of the new star identification in the case of Müla. Burgess identifies the *yogatārā* of this *nakṣatra* as Lambda Scorpionis, but several other authorities disagree. We can see from the figure that Lambda Scorpionis barely moves over a 50,000 year period. However, Epsilon Scorpionis moves almost directly towards the *jyotiṣa* position of Müla as we go back in time. Thus by visual inspection, it is a natural replacement for Lambda.

VII. Summary Discussion of the Three Hypotheses

According to the historical hypothesis, the 25,000 –55,000 year peak represents a period of astronomical measurement that took place roughly 25,000–55,000 years ago. Of course, the obvious objection to this is that at this time civilization didn't exist. Astronomy and the transmission of astronomical knowledge from generation to generation are thought to be impossible in this period.

It is generally accepted that human beings of modern type have existed for at least 40,000 years. But agriculture and settled village life are thought to have arisen only 7,000–10,000 years ago.[17] Early *Homo sapiens sapiens* presumably had the same talents and capabilities as modern humans, and one wonders why they waited 30,000–33,000 thousand years before starting the explosive development of civilized life. Could it be that civilization arose more gradually over a longer period of time?

Many ancient peoples apparently thought so. For example, fragments of the writings of Manetho, an Egyptian historian dated to the 3rd century B.C., give a period of some 30,000 years to early Egyptian civilization.[18] Berossus, a Babylonian historian of the same period, assigned 432,000 years to early Babylonian history,[19] and a similar time span is found in the king lists of Sumeria.[20] The period of 432,000 years is, of course, reminiscent of the Indian *yuga* cycles, even though these are regarded by some indologists as a recent invention by astronomers.[21]

It is easy to dismiss these traditions as sheer mythology, but it is possible that they may contain some truth. Perhaps sophisticated arts and sciences go back further than is generally believed.

Nonetheless, we may choose to reject the historical hypothesis. If we also reject the null hypothesis, then we are left with hypothesis three: There is a significant relationship between proper motions of stars and errors in *jyotiṣa* star positions, but this has nothing to do with the slow drift of stars after an ancient period of astronomical measurement. Unfortunately, since proper motions of stars are exceedingly slow, it is hard to see how their effects could become manifest over short periods of time to pre-telescopic astronomers. Long periods of time would seem to be necessary, and this suggests that some form of historical explanation is required.

I will therefore close by summing up the evidence presented in this paper that goes against the null hypothesis. This evidence can be summed up as follows:

1. The data for the 35 pairs of modern and *jyotiṣa* stars can be used to create an artificial distribution which has the same distributions of proper motions and error vectors as the 35 pairs, but eliminates all correlations between proper motions and error vectors. The null hypothesis predicts that the average angle of approach α for the 35 stars should be about the same as the average angle of approach μ for this artificial distribution. But actually α

17 Gowlett, John, Ascent to Civilization, Collins (London), 1984, pp. 120, 156.

18 Waddell, W. G., *Manetho*, Harvard Univ. Press (Cambridge), 1964.

19 Burstein, Stanley M., *The Babyloniaca of Berossus*, Undena Publications (Malibu), 1978.

20 Lambert, W. G. and Millard, A. R., Atrahasis: The Babylonian Story of the Flood, Clarendon Press (Oxford), 1969, p. 17.

21 Sethna, K. D., Ancient India in a New Light, Aditya Prakashan (New Delhi), 1989, p. 133.

is significantly lower than μ, as the historical hypothesis predicts. This also holds true if we restrict our analysis to the 28 *nakṣatras*.

2. We can formulate three sets of weighting factors based on historical studies of the *nakṣatras* having nothing to do with proper motions of stars. These weights have to do with Chinese *sieus*, the *nakṣatra* identifications of Alberuni, and *nakṣatra* identifications by a number of other scholars. When the average angle of approach α is calculated using these weights, it becomes lower and its statistical significance increases. According to the null hypothesis, this should not happen, but it can be readily explained in terms of the historical hypothesis.

3. The histogram of age estimates shows two pronounced peaks. The recent peak, corresponding to the period between 0 and 5,000 years ago, is made up of bright, prominent stars, and none of these stars correspond to Chinese *sieus*. According to the null hypothesis, it is quite improbable for this to happen. However, it is amenable to a simple historical explanation.

4. We can evaluate the statistical significance of the older histogram peak by looking at distances $D4_i$ of modern stars from *jyotiṣa* star positions 40,000 years ago, at a time corresponding to the middle of this peak. If we use the weights $Sieu_i$, which eliminate the recent peak, we find that these distances are significantly lower than would be expected by the null hypothesis. The level of significance is 1.5 standard deviations using $Sieu_i$ and 1.7 standard deviations using $Sieu_i \times Alb_i$.

5. If we visually examine graphs of star movement for the 28 *nakṣatras*, we can find five cases where the *yogatārā* chosen by Burgess does not move towards the *jyotiṣa* star position as we go back in time, but there is another star in the *nakṣatra* constellation which does this. All of the results reported in points 1– 4 were based on Burgess's star identifications. If we substitute the five new star identifications, we would naturally expect the average angles of approach to become smaller. This is because the new identifications were chosen with this in mind. However, we wouldn't necessarily expect the age estimates for these new identifications to reinforce the 25,000–55,000 year peak. There are not enough stars in the *nakṣatra* constellations to give us the opportunity to deliberately make such identifications. Nonetheless, the five new identifications all fall close to the 25,000 –55,000 year peak (and shift its center a bit closer to 55,000 years). This is also unexpected, given the null hypothesis.

These five points add up to a strong case against the null hypothesis, and they give some direct support to the historical hypothesis. I would suggest that it might therefore be worthwhile to seek further evidence of high cultural achievements in time periods going back tens of thousands of years.

Figures

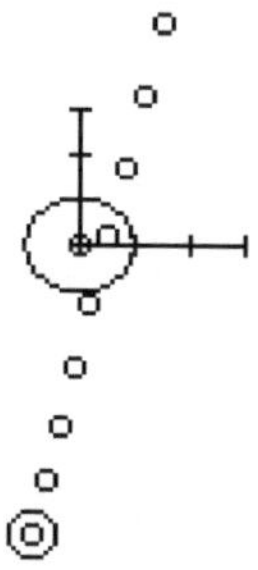

Figure 1. The large circle marks the *jyotiṣa* position of Brahmahṛdaya (see line 32 of Table 1). Burgess identifies this star as Capella (Alpha Aurigae).

The figure shows that Capella (double-ringed) was very close to the position of Brahmahṛdaya about 50,000 years ago. Its distance will be 6.487 degrees in A.D. 2000, and it reached a minimum value of 0.453 degrees in 48,164 B.E. (B.E. means before the epoch of A.D. 2000.)

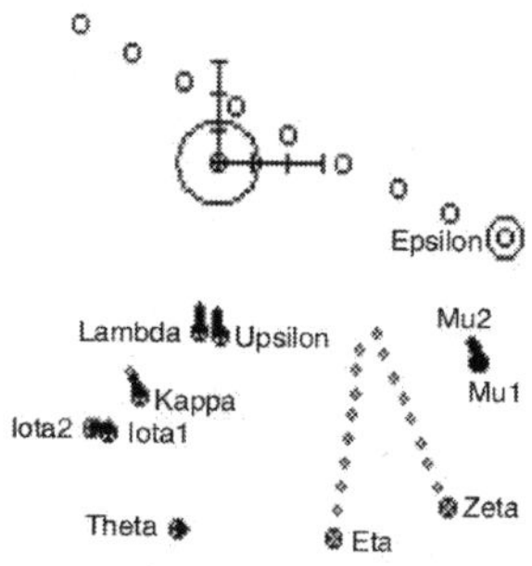

Figure 2. The large circle marks the *jyotiṣa* position of Mūla (see line 19 of Table 1). Colebrooke and Burgess agree that this *nakṣatra* should consist of the stars Upsilon Lambda Kappa Iota Theta Eta Zeta Mu Epsilon Scorpionis that form the tail of the scorpion.

The *Sūrya-siddhānta* says that the *yogatārā* is the eastern star of the group, and this is Iota. However, Colebrooke selects Upsilon and Burgress selects Lambda. This is done on the basis of position and agreement with the Arabic manzil ash-Shaulan, which consists of Upsilon and Lambda.

It turns out that Upsilon and Lambda barely move over a 50,000 year period. However, the figure shows that the star Epsilon reached a minimum distance of 1.667 degrees from the position of Mūla in 48,223 B.E. Therefore I have chosen Epsilon in Table 4 as a replacement for Burgess's choice of Lambda.

Tables

No.	Jyotiṣa star	Modern star	D	D_{min}	D4/D	A
1	Aśvinī	Beta Arietis	1.27	.58	.59	27.12
2	Bharaṇī	35 Arietis	1.29	1.08	.95	57.13
3	Kṛttikā	Eta Tauri	.61	.09	.19	8.75
4	Rohiṇī	Alpha Tauri	1.65	.24	.53	8.52
5	Mṛgaśirṣa	Lambda Orionis	4.40	2.87	.99	40.76
6	Ārdrā	Alpha Orionis	6.26	6.25	1.00	86.84
7	Punarvasu	Beta Geminorum	.69	.31	9.33	26.55
8	Puṣya	Delta Cancri	1.75	1.58	1.41	64.76
9	Āśleṣā	Epsilon Hydrae	5.23	5.21	1.06	85.34
10	Maghā	Alpha Leonis	.21	.15	12.32	43.31
11	Pūrvaphalgunī	Delta Leonis	3.77	3.77	1.51	151.18
12	Uttaraphalgunī	Beta Leonis	.41	.41	14.62	145.01
13	Hasta	Delta Corvi	2.84	.49	.17	9.89
14	Citrā	Alpha Virgini	.94	.58	.64	38.09
15	Svātī	Alpha Bootis	2.85	.43	7.25	8.64
16	Viśakhā	24 Iota Librae	2.59	2.23	.90	59.60
17	Anurādhā	Delta Scorpii	3.05	3.03	.99	82.29
18	Jyeṣṭhā	Alpha Scorpii	1.50	1.20	.90	53.07
19	Mūla	Lambda Scorpii	5.02	.33	.94	3.83
20	Pūrvaṣāḍhā	Delta Sagittari	1.25	1.25	1.22	111.21
21	Uttaraṣāḍhā	Sigma Sagittari	2.31	2.31	1.23	146.56
22	Abhijit	Alpha Lyri	1.68	.99	1.46	36.02
23	Śravaṇa	Alpha Aquilae	.71	.38	7.72	32.23
24	Dhaniṣṭhā	Beta Delphini	3.20	2.96	.92	67.58
25	Śatabhiṣaj	Lambda Aquarii	.67	.43	.65	40.10
26	Pūrvabhādrapadā	Alpha Pegasi	3.46	3.21	.94	68.34
27	Uttarabhādrapadā	Alpha Andromedo	5.78	5.78	1.09	92.21
28	Revatī	Zeta Piscium	1.18	1.18	2.15	124.16
29	Agastya	Alpha Carinae	.85	.11	.55	7.45
30	Mṛgavyādha	Alpha Can. Maj.	1.12	.99	11.13	62.61
31	Agni	Beta Tauri	7.92	7.12	.92	63.97
32	Brahmahṛdaya	Alpha Aurigae	6.49	.45	.20	4.01
33	Prajāpati	Delta Aurigae	6.57	2.50	.77	22.40
34	Apāmvatsa	Theta Virginis	1.84	1.16	.79	39.02
35	Āpas	Delta Virginis	5.91	3.29	.56	33.83

Table 1. Identification of modern stars corresponding to stars listed in the *Sūrya-siddhānta.* These identifications were given by Burgess. The quantities D, Dmin, D4/D, and A were calculated for this set of identifications. These quantities are defined in the text.

Nakṣatra	Modern star	4-Alt	Alb	Sieu	Sieu×Alb
Aśvinī	Beta Arietis	2	2	3	6
Bharaṇi	35 Arietis	1	1	3	3
Kṛttikā	Eta Tauri	3	2	3	6
Rohiṇī	Alpha Tauri	3	2	2	4
Mṛgaśirṣa	Lambda Orionis	2	2	3	6
Ārdrā	Alpha Orionis	1	0	0	0
Punarvasu	Beta Geminorum	3	2	0	0
Puṣya	Delta Cancri	3	2	3	6
Āśleṣā	Epsilon Hydrae	0	0	2	0
Maghā	Alpha Leonis	3	2	0	0
Pūrvaphalgunī	Delta Leonis	2	2	0	0
Uttaraphalgunī	Beta Leonis	3	2	0	0
Hasta	Delta Corvi	2	2	3	6
Citrā	Alpha Virgini	3	2	3	6
Svātī	Alpha Bootis	3	2	0	0
Viśakhā	24 Iota Librae	1	0	2	0
Anurādhā	Delta Scorpii	2	2	2	4
Jyeṣṭhā	Alpha Scorpii	3	2	2	4
Mūla	Lambda Scorpii	0	2	2	4
Pūrvaṣāḍhā	Delta Sagittari	2	2	1	2
Uttaraṣāḍhā	Sigma Sagittari	0	2	1	2
Abhijit	Alpha Lyri	3	2	0	0
Śravaṇa	Alpha Aquilae	3	2	0	0
Dhaniṣṭhā	Beta Delphini	2	0	0	0
Śatabhiṣaj	Lambda Aquarii	3	0	0	0
Pūrvabhādrapadā	Alpha Pegasi	3	0	3	0
Uttarabhādrapadā	Alpha Andromedo	2	0	2	0
Revatī	Zeta Piscium	2	0	0	0

Table 2. Weighting factors. These weights represent the degree of certainty of scholars in the identification of the *yogatārās* of the 28 *nakṣatras* (4-Alt and Alb) and the degree of correspondence between Chinese lunar mansions and *nakṣatras* (Sieu). The weight Sieu×Alb is simply the product of the weights Sieu and Alb.

	Weights	α	Mean	S.D.	No. of S.D.s
1	35 Ones	55.78	82.06	8.61	-3.05
2	28 Ones	61.40	82.43	9.67	-2.17
3	4-Alt_i	55.61	82.43	10.65	-2.52
4	Alb_i	54.69	82.43	11.51	-2.41
5	$Sieu_i$	49.30	82.43	12.92	-2.56
6	$Sieu_i \times Alb_i$	40.92	82.43	14.95	-2.78

Table 3. Weighted averages of the angles of approach. Here α is the weighted average of the Ai's using the indicated weights. In line 1 the average is for 35 stars with unit weights. In the remaining lines the average is limited to the 28 *nakṣatras*. The theoretical mean and standard deviation for 35×35 or 28×28 artificial stars are listed along with the number of standard deviations between α and the mean.

No.	Jyotiṣa star	Modern star	D	D_{min}	D4/D	A
2	Bharaṇī	41 Arietis	2.69	2.25	.84	56.67
5	Mṛgaśirṣa	Phi2 Orionis	5.16	.82	.31	9.20
8	Puṣya	Theta Cancri	1.82	.50	.53	16.05
11	Pūrvaphalgunī	Theta Leonis	1.88	.02	.38	.54
19	Mūla	Epsilon Scorpii	8.82	1.67	.25	10.93

Table 4. These star identifications are replacements for the corresponding identifications of Burgess in Table 1. Each replacement star belongs to the same *nakṣatra* constellation as Burgess's *yogatārā*, but it shows a stronger tendency to approach the *jyotiṣa* position as we go back in time.

Time in Vedic Culture as Presented in the Teachings of A. C. Bhaktivedanta Swami Prabhupāda

Christopher J. Hayton
Florida State University

Presented at the
Modern Journey Through Ancient Cosmography Seminar
November 21–22, 2020

Preface

The Hare Krishna Movement officially instituted by His Divine Grace A. C. Bhaktivedanta Swami Prabhupāda (henceforth Śrīla Prabhupāda, or just Prabhupāda) as the International Society for Krishna Consciousness (ISKCON)[1] appeared in the West in the mid-1960s. As a transplant or 'export' of traditional Gauḍīya (Bengali) Vaiṣṇava (Vaishnav) culture,[2] it has attracted scholarly attention since the early 1970s. While initially popularly thought of disparagingly[3] as a cult or new religious movement, it became increasingly apparent that the Hare Krishna Movement was indeed a genuine expression of the Vedic tradition of India, and specifically of the Gauḍīya Vaiṣṇava "Saṅkīrtana" Movement begun in the early sixteenth century by Caitanya Mahāprabhu.[4] Moreover, its authenticity became recognized by the global

1 Satsvarūpa dāsa Goswami, "The Biography of a Pure Devotee: Chartering A New Society," *Back to Godhead*, 15:5 (1980), 9–11.

2 John Y. Fenton, *Transplanting Religious Traditions. Asian Indians in America* (New York: Praeger, 1988), 144–147.

3 E. Burke Rochford, Jr., *Hare Krishna in America* (New Brunswick, NJ: Rutgers University Press, 1985), 269.

4 Thomas J. Hopkins, "The Social and Religious Background for Transmission of Gauḍīya Vaisnavism to the West," in *Krishna Consciousness in the West*, ed. David G. Bromley and Larry D. Shinn (Cranbury, NJ: Associated University Presses, 1989), 51

diasporic Hindu community,[5] which has, to an extent that is considerable in some locations, developed a symbiotic relationship with ISKCON,[6] thus helping to cement the identity of the Movement's adherents as a religious minority.[7]

Śrīla Prabhupāda brought with him a trunk when he arrived in Boston from India, and then onto New York by steam cargo ship in 1965.[8] The unpacking of that trunk has involved not just the physical removal and distribution of the books he printed in India and carried with him. A colorful, rich, complex culture has since been emerging, and in a very real sense that trunk is still being unpacked.

The beliefs and exotic appearance, practices, and lifestyle of the Hare Krishna devotees were naturally aspects of Gauḍīya Vaiṣṇava culture that generated initial interest in academia, investigations primarily coming from fields in the social sciences. The complete culture Prabhupāda represented as a living exponent is deep and complex in every way. An aspect of Gauḍīya Vaiṣṇava culture that so far has received relatively little consideration from Western academics outside the Hare Krishna Movement is its cosmology.

There is a history of modern Western astrophysics acknowledging the complexity and cyclical nature of time in the Vedic paradigm, and the similarity this generates with oscillating or cyclic models of the universe.[9] One aspect of this similarity is the enormity of the time scales involved, which parallel those deduced by modern science. This paper focuses on this one area of Vedic cosmology, specifically "time." Using translations and commentaries published by A. C. Bhaktivedanta Swami Prabhupāda during the 1960s and 70s as primary source material, details of the system of time present in the Vedic world view as it has come to us through the Gauḍīya Vaiṣṇava line are described. Key texts examined are the ancient Sanskrit texts of *Bhagavad-gītā*, *Śrīmad-Bhāgavatam* (the *Bhāgavata Purāṇa*), *Śrī Īśopaniṣad*, and *Brahma-saṁhitā*, as well as the medieval Bengali text, *Śrī Caitanya-caritāmṛta*. Expressed in culture-specific imagery that seemingly clashes with Western science, the accounts of the Vedic sages reach across time to reveal that the ancients grappled with

5 Richard J. Cole, "Forty Years of Chanting: A Study of the Hare Krishna Movement from its Foundation to the Present Day," in *The Hare Krishna Movement: Forty Years of Chant and Change*, ed. by Graham Dwyer and Richard J. Cole (London: I.B. Tauris & Co Ltd, 2007), 42–46.

6 E. Burke Rochford Jr., *Hare Krishna Transformed* (New York: New York University Press, 2007), 181.

7 Malory Nye, *Multiculturalism and Minority Religions in Britain: Krishna Consciousness, Religious Freedom, and the Politics of Location*, (Richmond, Surrey, UK: Curzon Press, 2001), 5.

8 Satsvarūpa dasa Goswami, *Srila Prabhupāda-līlāmṛta, Volume 2: Planting the Seed – New York City 1965–1966*, (Los Angeles, CA: The Bhaktivedanta Book Trust, 1980), 5–7.

9 Paul J. Steinhardt and Neil Turok, *Endless Universe: Beyond the Big Bang* (London: Weidenfeld & Nicolson, 2008), 169.

questions familiar to present day thinkers, and perhaps in some ways arrived at similar understandings. This is a living, extant system still guiding the lives of millions of people, including practitioners in the Hare Krishna Movement.

In attempting to try and understand Vedic (or Purāṇic) reality, it is necessary to put aside the conditioning acquired from modern society. An effective approach can, in some respects, be summarized by this statement from John S. Henderson in his book *The World of the Ancient Maya*:

> Maya reality, then, differs radically from most others... It is difficult for the outsider to appreciate these seemingly bizarre notions as a coherent, functional system of integrated beliefs. Every society believes that its own version of reality is "real" and classes others as fantasy – charming, perhaps, but surely distorted. To escape this trap, one must suspend disbelief to a degree that strains the rationalist mind. Yet only so is it possible to begin to achieve the empathy necessary to enter at least a little way into the Maya universe.[10]

Just as aspects of Maya thought appear to fundamentally contradict the viewpoints of Western science, so too does the world view present in Vedic culture. For the Vaiṣṇava, the Vedic worldview describes all reality, but for most Westerners, in whom conditioning in modern scientific theories runs deep, aspects of this reality can "go against the grain," particularly because the Vedic worldview is not restricted to objectivity as defined by Western science. As with Henderson's point regarding the Maya, in attempting to understand the Vedic worldview it is important to appreciate that it describes the origin and manifestation of the universe and its time component using scripts that feature the Supreme Entity and expansions thereof in a cosmic drama. In Vedic culture, these accounts are simultaneously stories featuring a variety of personalities and their functions and activities, and at the same time scientific treatises explaining the relationships between the forces these characters both represent and actually are, including the outcomes of their actions and interactions. This paper focuses on time in the context of this wider Vedic paradigm, offering a summary that describes the system of time as presented in the translations and accompanying commentaries or "purports" of the Vedic literatures written and published by A. C. Bhaktivedanta Swami Prabhupāda. The intention with this paper is to present this information without analysis.

10 John S. Henderson, *The World of the Ancient Maya* (Ithaca, NY: Cornell University Press, 1981), 73.

Introduction: Time in the Vedic Cosmos

In attempting to understand the Vedic world view and the place of time within it, it is fundamentally important to note the apparently dual nature of Vedic reality.[11] In the Vedic paradigm, everything that exists is part of the sum total of all existence,[12] an entity which is non-material, infinite,[13] eternal,[14] omnipresent,[15] and the source of everything we perceive as existing in the material realm[16, 17] as well as everything non-material or spiritual.[18] In the texts referred to in this paper, that entity is conscious and known as Kṛṣṇa (Krishna), Gauḍīya Vaiṣṇava theology being personalist and monotheistic. According to Vedic texts, part of that unlimited non-material entity transforms into material energy, from which a multitude of universes, including ours, manifest. As those texts also teach, everything material goes through a series of transformations, first coming into being, growing or developing, staying for some time, generating by-products, then dwindling and dying. In Vedic culture, this is just as true for the universe, stars, and planets, as it is for the bodies of living beings.[19] Most importantly, these material manifestations are temporary, or subject to the influence of time. As Kṛṣṇa states in *Bhagavad-gītā*, "Time I am, the great destroyer of the worlds."[20] The dual nature of existence is represented by, on the one hand, an unlimited spiritual nature (*paravyoma*,

---

11 A. C. Bhaktivedanta Swami Prabhupāda, *Bhagavad-gītā As It Is* 8.20 (New York: Collier Books, 1972), 430: "Yet there is another nature, which is eternal and is transcendental to this manifested and unmanifested matter. It is supreme and is never annihilated. When all in this world is annihilated, that part remains as it is."

12 A. C. Bhaktivedanta Swami Prabhupāda, *Śrī Īśopaniṣad* (Los Angeles, CA: ISKCON Books, 1969), 11: "Because He is the Complete Whole, even though so many complete units emanate from Him, He remains the complete balance." [Invocation].

13 A. C. Bhaktivedanta Swami Prabhupāda, *Śrīmad-Bhāgavatam, The One Volume Edition* 2.6.40 (Manila, Philippines: The Bhaktivedanta Book Trust, 1982), 334: "He is all-pervading, without beginning or end ..."

14 Prabhupāda, *Bhagavad-gītā* 2.12, 86: "Never was there a time when I did not exist ..."

15 Prabhupāda, *Bhagavad-gītā* 9.4, 503: "By Me, in My unmanifested form, this entire universe is pervaded."

16 Prabhupāda, *Sri Īśopaniṣad*, 11 [Invocation].

17 Prabhupāda, *Bhagavad-gītā* 7.6, 370: "Of all that is material and all that is spiritual in this world, know for certain that I am both its origin and dissolution."

18 Prabhupāda, *Bhagavad-gītā*, 10.8, 450: "I am the source of all spiritual and material worlds. Everything emanates from Me."

19 A. C. Bhaktivedanta Swami Prabhupāda, '760102 - Lecture SB 07.06.01 – Madras', *Vanisource* (2022) <https://vanisource.org/wiki/760102_-_Lecture_SB_07.06.01_-_Madras> [accessed August 15, 2023]: "This material world, it will not exist forever. It is *bhūtvā bhūtvā pralīyate*. It is manifested at a certain date. Just like your body and my body, it is manifested on a certain date; it will stay for some time; it will grow; it will give some by-product; then we become old, dwindling, and then finished. This is called *ṣaḍ-vikāra*. Of anything which is material. But there is another nature where there is no *ṣaḍ-vikāra*. That is eternal."

20 Prabhupāda, *Bhagavad-gītā*, 11.32, 563.

translated as "spiritual sky"[21]), and on the other by a portion of that spiritual nature transformed into material nature, from which, in combination with the living entities, minute fragmental parts of the Supreme Consciousness, the total material energy, known in Sanskrit as the *mahat-tattva*, is produced.[22] Spiritual and material nature therefore constitute this duality. The material cosmos, which emerges from this *mahat-tattva*, is alternately unmanifest and manifest in a never-ending cycle,[23] and as the 10th Century South Indian Vaiṣṇava devotee, Śrī Yāmunācārya, observed, "...everything within material nature is limited by time, space and thought."[24]

Key to understanding the place of time in the Vedic paradigm are the Viṣṇu expansions of the Supreme Consciousness, of the Supreme Entity, Kṛṣṇa. For the purpose of creation or manifestation of the material cosmos, the Sum Total of All Existence (Kṛṣṇa) assumes particular forms, in this case called in Sanskrit *puruṣa-avatāras*, with all the necessary energy and potency to carry out this function. These are the three forms of Viṣṇu involved in the "flowering" or unfolding of the material cosmos. The third Viṣṇu expands from the second, and the second from the first. That first Viṣṇu is known as Mahā-Viṣṇu, or Kāraṇodakaśāyī Viṣṇu, the One who lies in the Causal (Kāraṇa) Ocean[25] in a state of mystic slumber (*yoga-nidrā*). The Causal Ocean is entirely spiritual,[26] and it is this first *puruṣa* incarnation, Mahā-Viṣṇu, who creates the total material energy (*mahat-tattva*).[27]

It is the breathing of the eternally sleeping Mahā-Viṣṇu that begins the introduction of time as it is generally understood. As Mahā-Viṣṇu breathes out, innumerable universes emanate from the pores of His transcendental body, and these universes re-enter His body as He breathes in again. During

21 A. C. Bhaktivedanta Swami Prabhupāda, *Sri Caitanya-caritāmṛta: The One Volume Edition, Adi-līlā* (Singapore, Malaysia: Gauranga PTE Ltd., 1984), 5.14, 145: "Beyond the material nature lies the realm known as *paravyoma*, the spiritual sky."

22 Prabhupāda, *Caitanya-caritāmṛta, Madhya-līlā* 20.274, 1505–1506: "From time immemorial, after agitating the material nature into three qualities, the Supreme Personality of Godhead placed the semen of innumerable living entities within the womb of that material nature. Thus, material nature gives birth to the total material energy, known as the *hiraṇmaya-mahat-tattva*, the original symbolic representation of the cosmic manifestation."

23 Prabhupāda, *Caitanya-caritāmṛta, Adi-līlā* 1.53 purport, 17: "The material cosmos, being temporary, is sometimes manifest and sometimes unmanifest..."

24 Kuśakratha dāsa and Pūrṇaprajña dāsa, *Śrī Stotra-ratna: Śrīla Yāmunācārya* (Vrindaban, India: Rasbihari Lal & Sons, 2014), p. 16 [Text 13]. The entirety of *Śrī Stotra-ratna* verse 13 is incorporated into Prabhupāda, *Caitanya-caritāmṛta, Antya līlā*, 3.02.

25 Prabhupāda, *Caitanya-caritāmṛta, Adi-līlā* 5.50, 175: "... Kāraṇodakaśāyī Viṣṇu, lying on the Kāraṇa Ocean, is the original *puruṣa*..."

26 Prabhupāda, *Caitanya-caritāmṛta, Adi-līlā* 5.54, 177: "The water of the Kāraṇa Ocean, which is the original cause, is therefore spiritual."

27 Prabhupāda, *Caitanya-caritāmṛta, Adi-līlā* 5.56, 177: "He is known as the first *puruṣa*, the creator of the total material energy."

each breath universes come into being and are subsequently destroyed.[28] It is this alternating state that has drawn comparisons with the oscillating universe theories of modern physics that involve a cyclical "Big Bang" and "Big Crunch."

As the universes come into being, Mahā-Viṣṇu expands into each universe as Garbhodakaśāyī Viṣṇu. This is summarized by Śrīla Prabhupāda as follows:

> The second Viṣṇu incarnation, the Garbhodakaśāyī Viṣṇu, enters each and every universe, spreads water from His body, and lies down on that water. From His navel, the stem of a lotus flower grows, and on that lotus flower the first creature, Brahmā, is born. Within the stem of that lotus flower are fourteen divisions of planetary systems, which are created by Brahmā. [29]

This image is a core component of Vaiṣṇava culture, and in some temples, it is directly this form of Viṣṇu that is worshiped, for example, in the famous Padmanābhaswamy (Ananta Padmanābha Viṣṇu) Temple in Thiruvananthapuram (Trivandrum) in Kerala State, one of the temples visited by Caitanya Mahāprabhu on His tour of South India.[30] Brahmā is the secondary creator, tasked with providing material bodies for the population of souls that have entered the universe, and planets for them to reside on. The planetary systems mentioned in the quote above are planes of existence within a tiered universe, where upper planetary systems are often considered to correspond to higher states of consciousness, and become progressively more covered by ignorance in the lower systems. The earthly realm is a middle stratum. All are subject to the time factor, as is described in *Bhagavad-gītā*: "From the highest planet in the material world down to the lowest, all are places of misery wherein repeated birth and death take place."[31] Significantly, Brahmā lives for the duration of the universe and his life span is intimately tied with the life of the universe. He is described as living for one hundred of his years, each year being made up of twelve of his months, 360 of his days, etc. Putting this in a human perspective, of Brahmā, Śrīla Prabhupāda writes, "his one day is calculated at 4,300,000,000 of our earthly years. His night is of the same duration."[32] Situated at the upper end of the Vedic time scale as it applies to the material

---

28 A. C. Bhaktivedanta Swami Prabhupāda, *Teachings of Lord Chaitanya: A Treatise on Factual Spiritual Life* (London: Bhaktivedanta Book Trust, 1985), 86: "As innumerable particles of dust pass through the tiny holes in a screen, similarly from the pores of the Mahā-Viṣṇu's body innumerable universes emanate. As He breathes out, innumerable universes are produced, and as He inhales, they are annihilated."

29 Prabhupāda, *Teachings of Lord Chaitanya*, 86.

30 Prabhupāda, *Caitanya-caritāmṛta, Madhya-līlā* 1.115, 819: "Śrī Caitanya Mahāprabhu then visited the Viṣṇu temples of Anantadeva, Puruṣottama, Śrī Janārdana, Padmanābha and Vāsudeva."

31 Prabhupāda, *Bhagavad-gītā* 8.16, 426.

32 Prabhupāda, *Bhagavad-gītā* 9.7, 455.

world, therefore, is the lifetime of Brahmā, representing the duration of the universe, and then the subdivisions of Brahmā's lifespan.

As everything within the material universe is manifested, Garbhodakaśāyī Viṣṇu expands as Kṣīrodakaśāyī Viṣṇu. This form of Viṣṇu residing within every atom and alongside each embodied living being is translated by Śrīla Prabhupāda as the Supersoul, Paramātmā, meaning the Supreme Consciousness individually present everywhere. Prabhupāda writes: "Another name of the Paramātmā feature of the Lord is *kāla*, or eternal time. Eternal time is the witness of all our actions, good and bad, and thus resultant reactions are destined by Him."[33] The Tenth Canto of *Śrīmad-Bhāgavatam* describes time as "another form of Lord Viṣṇu."[34]

The material cosmos and the universes within it constitute a stage upon which a cosmic drama is played out, with a grand sequence of timed events in a general plot that begins with the universes coming into being and ends with their destruction. The actors in this play are the fragmental parts and parcels of the Supreme Entity, the *jīvas,* or individual conscious living entities (souls), who, because of having free will, are ad-libbing their way through the scenes in situations determined by the structure of sets that accommodate their karma and desires. The Paramātmā, as Time, facilitates all that is taking place in this moment-by-moment simultaneous writing and enacting of the script. When one season is over, another begins[35] in this ever-popular material production, in which the audience are also the actors, and the individuals taking part come and go, depending on their earned qualifications and how they utilize their experiences.

Divisions of Time

Structure of the Universe

Śrīmad-Bhāgavatam describes the universe as a multi-tiered, multi-dimensional environment wherein the living beings experience time differently according to the level of consciousness in which they are situated. Śrīla Prabhupāda explains, "Planets are of different types, each controlled by a different *kāla-cakra,* or schedule of eternal time. Time on Earth is not necessarily applicable to other planets, which may be located in a more elevated

33 Prabhupāda, *Śrīmad-Bhāgavatam* 1.8.28, purport, 92.

34 A. C. Bhaktivedanta Swami Prabhupāda, *Śrīmad-Bhāgavatam: Tenth Canto-Part One,* (Los Angeles, CA: Bhaktivedanta Book Trust, 1977), 10.3.26, 238.

35 Prabhupāda, *Śrīmad-Bhāgavatam* 3.10.13, purport, 512: "There is a systematic schedule for the perpetual manifestation, maintenance and annihilation of the material world ..."

stratum."[36] In such higher realms time is more dilated than in lower realms. Nevertheless, the length of lifespan of different species of life is experienced by the entities within those forms as relatively the same in, for example, the ant and the human and the elephant, although actually the lifespans are of different lengths from our perception. Brahmā's day of 4,300,000,000 earth years is experienced by him as a day, just as a day is experienced by us in our stratum. Prabhupāda explains the time dilation effect: "In the upper planetary system the duration of one complete day and night is equal to one complete year of this earth."[37]

Śrīmad-Bhāgavatam gives an example that illustrates the time dilation effect that would be experienced by persons able to move from a lower to a higher stratum within the universe. Wanting to find a husband for his daughter, Revatī, King Kakudmī took her to see Lord Brahmā in Brahmaloka, the highest planet in the material cosmos. When they arrived there, Lord Brahmā was busy listening to Gandharvas (inhabitants of the heavenly planets) performing musical arrangements, and wasn't able to speak with him immediately. Kakudmī waited until the performance was finished, then after offering respects to Lord Brahmā, explained the reason for his visit. Lord Brahmā laughed and told the King that while he had been waiting, so many generations of descendants of those whom he had been considering as potential sons-in-law had passed away that their names could no longer be remembered. A short while in Brahmaloka had been equivalent to a vast expanse of time in the lower realm of the universe from whence King Kakudmī had come.[38]

Cyclical Time[39]

Ancient Vedic culture possesses a cyclical concept of time, each cycle having subdivisions which are themselves cycles (see Figs. 1–5). Prabhupāda explains that in the Vedic paradigm, during the process of the manifestation of the material universe "there are three different types of creation, called *mahā-kalpa, vikalpa* and *kalpa*, of which "there are regular and systematic cycles."

Mahā-kalpa

The *Mahā-kalpa* constitutes creations by Kāraṇodakaśāyī Viṣṇu (Mahā-Viṣṇu). The exhalations and inhalations of Viṣṇu are *mahā-kalpas*, these breaths

36 Prabhupāda, *Śrīmad-Bhāgavatam* 3.11.16, purport, 518.

37 Prabhupāda, *Śrīmad-Bhāgavatam* 2.10.46, purport, 405.

38 Prabhupāda, *Śrīmad-Bhāgavatam* 9.3.29–32, 2064.

39 Note: Most of the information in this and the following three sections comes from Prabhupāda, *Śrīmad-Bhāgavatam* 2.10.46, purport, 405–6. Other sources will be referenced individually.

making a *vikalpa* possible, in which Brahmā, the secondary creator, is created as the incarnation of spiritual sound.[40] Mahā-Viṣṇu is entirely spiritual, so the breathing of Mahā-Viṣṇu is not material and therefore not under the influence of time, yet as it takes place, the material energy reposed in Mahā-Viṣṇu becomes transformed into the innumerable material universes emanating from the pores of His body. It is in the next phase of creation that the time element becomes measurable. It should be noted here that the word "universe" as used by Śrīla Prabhupāda in his translations, purports, and lectures, often refers to the Sanskrit word *brahmāṇḍa*:

> Brahmā means the manager of one *brahmāṇḍa*, universe. And there are millions of *Brahmās*, millions, trillions. They are numberless. *Ananta-koṭi. Yasya prabhā prabhavato jagad-aṇḍa-koṭiḥ.*[41] *Koṭi. Koṭi* means unlimited. *Jagad-aṇḍa. Jagad-aṇḍa* means universe. *Brahmāṇḍa* or *jagad-aṇḍa. Aṇḍa. Aṇḍa* means it is egg-shaped, round, egg-shaped. Therefore, it is called *aṇḍa, brahmāṇḍa.*[42]

Vikalpa

"The creation of Brahmā and dispersion of the material ingredients are called *vikalpa.*" *Vikalpas* are generated by the breathing of Mahā-Viṣṇu, as stated in the *Brahma-saṁhitā*: "The Brahmās live only during the breathing period of Mahā-Viṣṇu"[43] (Fig. 1). According to *Bhagavad-gītā*, as Brahmā's day begins, the "multitude of living entities comes into being, and at the arrival of Brahmā's night they are all annihilated."[44] Brahmā's life consists of one hundred of his years,[45] each composed of twelve of his months. Each month of Brahmā is made up of thirty *kalpas,* or Days of Brahmā, each of which has a specific name. Figure 2 shows each of the thirty *kalpas* in a month of Brahmā, with the name of each *kalpa*.

---

40 Prabhupāda, *Śrīmad-Bhāgavatam* 3.12.47 purport, 533.

41 Bhaktisiddhānta Sarasvatī Gosvāmī Ṭhākura, *Sri Brahma-saṁhitā* (Mumbai, India: Bhaktivedanta Book Trust 2016), 5.40, 86.

42 A. C. Bhaktivedanta Swami Prabhupāda, '730730 - Lecture BG 01.43 – London', *Vanisource* (2020) <https://vanisource.org/wiki/730730_-_Lecture_BG_01.43_-_London?hl=brahmanda> [accessed August 15, 2023]

43 Bhaktisiddhānta, *Brahma-saṁhitā* 5.48, 102: "Brahmā and other lords of the mundane worlds, appearing from the pores of hair of Mahā-Viṣṇu, remain alive as long as the duration of one exhalation of the latter..."

44 Prabhupāda, *Bhagavad-gītā* 8.18, 429: "When Brahmā's day is manifest, this multitude of living entities comes into being, and at the arrival of Brahmā's night they are all annihilated."

45 Prabhupāda, *Śrīmad-Bhāgavatam* 1.3.5, 26.

Kalpa

"The creation by Brahmā in each day of his life is called *kalpa*..." (Fig. 3) and "therefore, each day of Brahmā is called a *kalpa*." Each *kalpa* ends with a partial devastation that affects "the lower planetary systems up to Svargaloka."[46] This is the night of Brahmā, during which he goes to sleep. During Brahmā's night, the three planetary systems below Brahmaloka (Svarga, Martya, and Pātāla) become "submerged in the water of devastation. In his sleeping condition, Brahmā dreams about the Garbhodakaśāyī Viṣṇu and takes instruction from Him for the rehabilitation of the devastated area of space."[47]

> After the end of Brahmā's night, the creation of the three worlds begins again in the daytime of Brahmā, and they continue to exist through the life durations of fourteen consecutive Manus, or fathers of mankind. At the end of the life of each Manu there are shorter dissolutions also.[48]

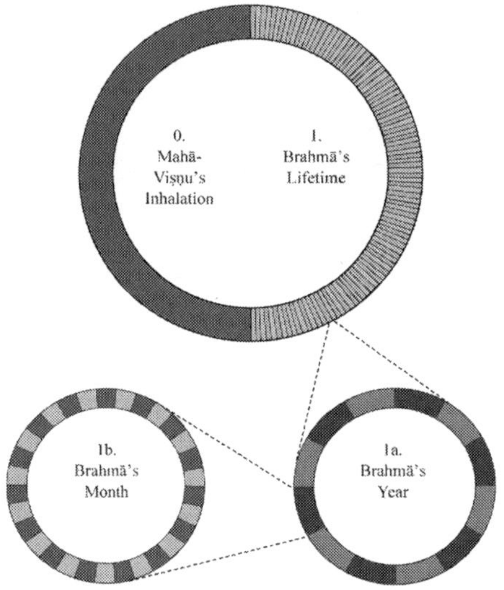

Figure 1: Diagram Illustrating the Life of Lord Brahmā.
0. The material cosmos is wound up in the body of Mahā-Viṣṇu during Mahā-Viṣṇu's inhalation.

46 A. C. Bhaktivedanta Swami Prabhupāda, *KṚṢṆA* (London: Bhaktivedanta Book Trust, 1986), Ch. 81 ("Lord Kṛṣṇa and Balarāma Meet the Inhabitants of Vṛndāvana") 693.

47 Prabhupāda, *Śrīmad-Bhāgavatam* 3.11.22, purport, 519.

48 Prabhupāda, *Śrīmad-Bhāgavatam* 3.11.23, verse and purport, 519.

1. Lord Brahmā lives during the exhalation of Mahā-Viṣṇu. His life consists of 100 of his years.
1a. A year of Lord Brahmā consists of twelve of his months.
1b. A month of Lord Brahmā consists of thirty of his days.

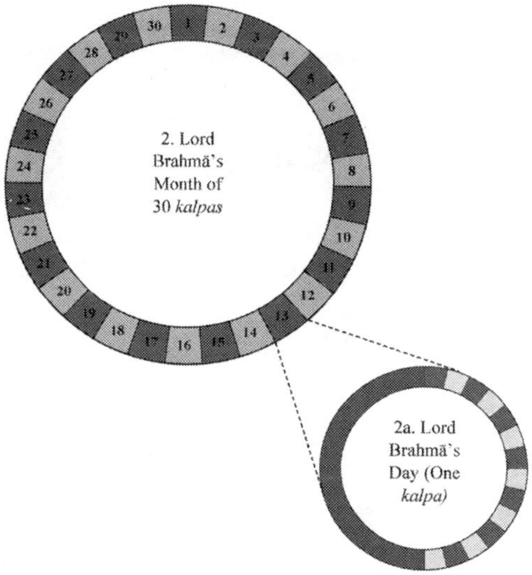

Figure 2: Diagram Illustrating a Month of Lord Brahmā.
Lord Brahmā's lifetime consists of 100 of his years, each of which is made up of twelve months each of thirty *kalpas*, or Days of Brahmā. The names of the thirty *kalpas* numbered in Figure 2 are as follows[49]:

1	Śveta-kalpa	11	Dhyāna	21	Somākalpa
2	Nīlalohita	12	Sārasvata	22	Bhāvana
3	Vāmadeva	13	Udāna	23	Supuma
4	Gāthāntara	14	Garuḍa	24	Vaikuṇṭha
5	Raurava	15	Kaurma	25	Arciṣa
6	Prāṇa	16	Nārasiṁha	26	Valīkalpa
7	Bṛhat-kalpa	17	Samādhi	27	Vairāja
8	Kandarpa	18	Āgneya	28	Gaurī-kalpa
9	Sadyotha	19	Viṣṇuja	29	Māheśvara
10	Īśāna	20	Saura	30	Paitṛ-kalpa

49 Prabhupāda, *Śrīmad-Bhāgavatam* 2.10.46 purport, 406.

The Day and Night of Brahmā

During Brahmā's daytime, all planetary systems are present and functional, and their inhabitants come and go based on their actions and reactions.[50]

In the tiered Vedic universe, the higher planetary systems are populated by demigods (*devas*), as are the highest systems. As each *kalpa* ends, it is followed by a partial devastation that affects all but the highest planetary systems. Rudra, or Śiva (Shiva, the Destroyer), "represented by the fire of eternal time," dissolves the three worlds:

> These three worlds are known as Bhūḥ, Bhuvaḥ and Svaḥ (Pātāla, Martya and Svarga). The innumerable living entities merge into that dissolution, which appears to be the dropping of the curtain of the scene of the Supreme Lord's energy, and so everything becomes silent.[51]

At the end of Brahmā's day, when the universe merges into the darkness of night, the blazing fire resulting in this partial devastation is said to rage

> for one hundred years of the demigods, or 36,000 human years. Then for another 36,000 years there are torrents of rain, accompanied by violent winds and waves, and the seas and oceans overflow. These reactions of 72,000 years are the beginning of the partial devastation of the three worlds.[52]

The Vedic cycle of universal creation and destruction therefore includes smaller devastations involving part but not all of the universe, the night of Brahmā being the largest of these that do not result in the total demise of the universe. Brahmā's daytime is subdivided into fourteen *manvantaras*, each with a specifically named *manvantara-avatāra*, or Manu incarnation, as illustrated in Figure 3. Information in Figure 3 regarding the fourteen Manus in a day of Brahmā is derived from the Eighth Canto of *Śrīmad-Bhāgavatam*, chapters 1, 5, and 13.[53]

---

50 Prabhupāda, *Śrīmad-Bhāgavatam* 3.11.26, 520: "In the creation, during Brahmā's day, the three planetary systems—Svarga, Martya and Pātāla—revolve, and the inhabitants, including the lower animals, human beings, demigods and Pitās, appear and disappear in terms of their fruitive activities."

51 Prabhupāda, *Śrīmad-Bhāgavatam* 3.11.28 purport, 520.

52 Prabhupāda, *Śrīmad-Bhāgavatam* 3.11.31, purport, 521.

53 Prabhupāda, *Śrīmad-Bhāgavatam*, 8.1.1–30, 1847–1855; 8.5.1–10, 1880–1882; 8.13.1–36, 1956–1962.

Sub-Divisions of a Kalpa

The Vedic time system contains a predictable large-scale sequence of events. The stories of individual lives occur within the parameters of this structure. As has been discussed above, the universe persists for a limited amount of time, and "is manifested in cycles of *kalpas*."[54] A *kalpa* is a day (12 hours) of Brahmā:

> The four *yugas* are calculated in terms of the heavenly calendars and accordingly are twelve thousand years in terms of the heavenly planets. This is called a *divya-yuga*, and one thousand *divya-yugas* make one day of Brahmā.[55]

> By human calculation, a thousand ages taken together is the duration of Brahmā's one day. And such also is the duration of his night.[56]

Prabhupāda further explains in his purport to this verse:

> The duration of the material universe is limited. It is manifested in cycles of *kalpas*. A *kalpa* is a day of Brahmā, and one day of Brahmā consists of a thousand cycles of four *yugas* or ages: Satya, Tretā, Dvāpara, and Kali.[57]

During one *kalpa*, there are 14 *manvantaras*, and 14 Manus appear. During Brahmā's life of one hundred of his years; there are 5,040 Manus in each year and a lifetime total of 504,000. This is the calculation for one universe, and "there are innumerable universes."[58] Prabhupāda uses the term *manvantara-avatāras* to include both the Manus themselves as well as a specific incarnation of Kṛṣṇa for each *manvantara*. In each *manvantara* there is also a different Indra (the demigod who is the King of the heavenly planets). Table 1 (next page) lists the 14 *manvantaras* and the corresponding identity of Indra for each.

*Note that the 11th and 12th *manvantaras* are given in a different order in Śrīla Prabhupāda's translations and purports. The list in Table 1 follows the order given in *Śrīmad-Bhāgavatam* 8.13. 24 and 25, and this order is corroborated in Prabhupāda's purport to *Caitanya-caritāmṛta, Adi-līlā* 3.9 and in *Caitanya-caritāmṛta, Madhya-līlā* 20.326 and 327. The alternate order for

54 Prabhupāda, *Bhagavad-gītā* 8.17, purport,, 428.
55 Prabhupāda, *Śrīmad-Bhāgavatam* 2.10.46, purport, 405.
56 Prabhupāda, *Bhagavad-gītā* 8.17.
57 Prabhupāda, *Bhagavad-gītā* 8.17 purport, 428.
58 Prabhupāda, *Śrīmad-Bhāgavatam* 1.3.5 purport, 26.

Table 1. The names of the fourteen *manvantaras* and their corresponding Indras during one *kalpa*.

	Manvantara	Indra
1.	Svāyambhuva	Yajña
2.	Svārociṣa	Rocana
3.	Uttama	Satyajit
4.	Tāmasa	Triśikha
5.	Raivata	Vibhu
6.	Cākṣuṣa	Mantradruma
7.	Vaivasvata	Purandara
8.	Sāvarṇi	Bali, son of Virocana
9.	Dakṣa-sāvarṇi	Adbhuta
10.	Brahma-sāvarṇi	Śambhu
11.	Dharma-sāvarṇi*	Vaidhṛta
12.	Rudra-sāvarṇi*	Ṛtadhāmā
13.	Deva-sāvarṇi	Divaspati
14.	Indra-sāvarṇi	Śuci

the 11th and 12th *manvantaras* is given in the purport to *Caitanya-caritāmṛta, Madhya-līlā* 20.328, where Prabhupāda refers to Śrīla Bhaktisiddhānta Sarasvatī Ṭhākura's *Anubhāṣya* as he is listing the fourteen Manus: "(11–14) Rudra-sāvarṇi, Dharma-sāvarṇi, Deva-sāvarṇi and Indra-sāvarṇi, the sons of Rudra, Ruci, Satyasahā and Bhūti respectively."[59]

The sequence of fourteen *manvantaras* repeats itself every day of Brahmā, although the living entities (spirit souls) taking on these identities and roles are different. Prabhupāda explains our current location in this sequence, using the word "millennium" to mean one cycle of the four *yugas* (see Figure 5) as follows:

> In one day of Brahmā there are fourteen Manus, who each live for seventy-one millenniums. The four *yugas*—Satya, Tretā, Dvāpara and Kali—constitute one millennium. We are presently in the *manvantara* of Vaivasvata Manu ... in the twenty-eighth millennium of Vaivasvata Manu ... Since each millennium consists of 4,300,000 years and

59 Prabhupāda, *Caitanya-caritāmṛta, Madhya-līlā* 20.238, 1514-1515.

> we are now in the twenty-eighth millennium, some 120,400,000 years have passed since ... the beginning of the Vaivasvata *manvantara*.[60]

According to this statement, we are about half way through the current day

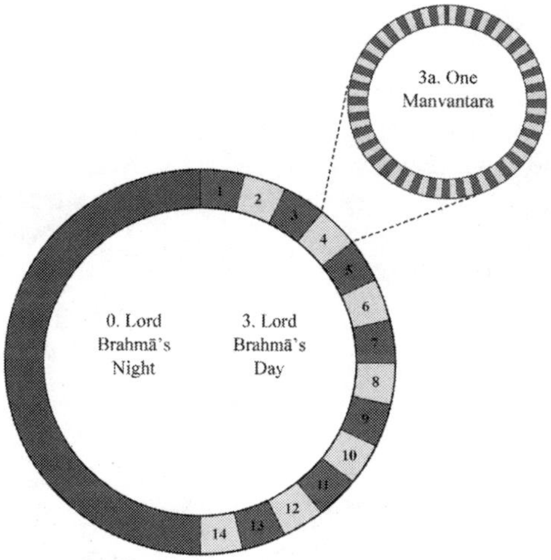

Figure 3: Diagram Illustrating a 24-Hour Day of Brahmā.

The twelve-hour day of Brahmā is called a *kalpa*. Brahmā's night is of the same length. In Figure 2, the *sandhyās*, or junctions, between each *manvantara* are not shown in the diagram.

A *kalpa* (day of Brahmā) lasts for 4.32 billion solar years. During one *kalpa* there are 14 *manvantaras* and 15 *sandhyās*, or junctions, preceding and following each *manvantara*. During the *sandhyās* a partial devastation and corresponding partial creation takes place within the universe. Each *sandhyā* lasts 1,728,000 solar years, the length of a Satya-yuga.

Within each *manvantara* there are 71 *divya-yugas* (cycles of the four *yugas*). The names of each *manvantara* numbered on the diagram in Figure 3 are as follows:

1	Svāyambhuva-manvantara	8	Sāvarṇya-manvantara
2	Svārociṣa-manvantara	9	Dakṣa-sāvarṇya-manvantara
3	Uttama-manvantara	10	Brahma-sāvarṇya-manvantara
4	Tāmasa-manvantara	11	Rudra-sāvarṇya-manvantara
5	Raivata-manvantara	12	Dharma-sāvarṇya-manvantara
6	Cākṣuṣa-manvantara	13	Deva-sāvarṇya-manvantara
7	Vaivasvata-manvantara	14	Indra-sāvarṇya-manvantara

60 Prabhupāda, *Śrīmad-Bhāgavatam* 6.10.16 purport, 1539.

of Brahmā. The *Śrīmad-Bhāgavatam* states that we are also about half way through Brahmā's life,[61] meaning that this universe is also half way through its lifespan. Tables 1 and 2 both illustrate the "past, present, and future" feature of the cosmic framework of Vedic time, with some *manvantaras* already having passed, the current *manvantara* under Vaivasvata Manu, and those scheduled to happen at a later time.

Table 2. The names of one set of fourteen Manus appearing during one day of the life of Brahmā within this universe, and the *avatāras* of the Supreme Lord appearing in the fourteen successive *manvantaras*.

#	Manu[62]	Father of Manu[63]	Incarnation of the Supreme Lord[64, 65]
1	Svāyambhuva	Lord Brahmā	Yajña*
2	Svārociṣa	Svārociḥ (Agni)	Vibhu
3	Uttama	King Priyavrata	Satyasena
4	Tāmasa	King Priyavrata	Hari
5	Raivata	King Priyavrata	Vaikuṇṭha
6	Cākṣuṣa	The demigod Cakṣu	Ajita
7	Vaivasvata (Śrāddhadeva)	Vivasvān, the sun god, and his wife Saṁjñā	Vāmana*
8	Sāvarṇi	Vivasvān, the sun-god, and his wife Chāyā	Sārvabhauma
9	Dakṣa-sāvarṇi	The demigod Varuṇa	Ṛṣabha
10	Brahma-sāvarṇi	Upaśloka	Viṣvaksena
11	Dharma-sāvarṇi	Ruci	Dharmasetu
12	Rudraputra (Rudra-sāvarṇi)	Rudra	Sudhāmā
13	Raucya or Deva-sāvarṇi	Satyasahā	Yogeśvara
14	Bhautyaka or Indra-sāvarṇi	Bhūti	Bṛhadbhānu

***Both Yajña and Vāmana are also counted among the *līlā-avatāras* (the incarnations meant for the performance of pastimes); Yajña also takes the post of Indra during the Svāyambhuva *manvantara*.[66]**

61 Prabhupāda, *Śrīmad-Bhāgavatam* 3.11.34. 521: "The one hundred years of Brahmā's life are divided into two parts, the first half and the second half. The first half of the duration of Brahmā's life is already over, and the second half is now current."

62 Prabhupāda, *Caitanya-caritāmṛta, Madhya-līlā* 20.328 purport, 1514–1515.

63 Prabhupāda, *Caitanya-caritāmṛta, Madhya-līlā* 20.328 purport, 1514–1515.

64 Prabhupāda, *Caitanya-caritāmṛta, Madhya-līlā* 20.246 purport, 1498.

65 Prabhupāda, *Caitanya-caritāmṛta, Madhya-līlā* 20.325–328, 1514.

66 Prabhupāda, *Caitanya-caritāmṛta, Madhya-līlā* 20.246, purport, 1498.

The Manu incarnations can also be referred to as *vaibhava-avatāras*.[67] Prabhupāda writes that according to Śrī Jīva Gosvāmī, supported by Śrīla Viśvanātha Cakravartī, "there is always a devastation after the end of each and every Manu."[68] The *Śrīmad-Bhāgavatam* further narrates:

> After the dissolution of each and every Manu, the next Manu comes in order, along with his descendants, who rule over the different planets; but the seven famous sages, and demigods like Indra and their followers, such as the Gandharvas, all appear simultaneously with Manu.[69]

Prabhupāda explains: "The duration of life of a Manu ..." (one *manvantara*, see Figure 4) "comprises seventy-one sets of four millenniums"[70] (using the word 'millennium' to translate the term *divya-yuga*). A *divya-yuga* is calculated in years of the demigods of the upper planetary system, where the duration of one complete day and night is equal to one complete year on Earth,[71] one demigod year being equal to 360 earth years.[72]

One *divya-yuga* (*catur-yuga*, cycle of the four ages: Satya, Dvāpara, Tretā, and Kali, including transitional *yuga-sandhyās*) = 4,320,000 solar years (12,000 demigod years[73]); each age is itself divided into 3 *sandhyās*. *Śrīmad-Bhāgavatam* 3.11.19 and Prabhupāda's purport to this verse provide details:

> The duration of the Satya millennium equals 4,800 years of the demigods; the duration of the Tretā millennium equals 3600 years of the demigods; the duration of the Dvāpara millennium equals 2,400 years; and that of the Kali millennium is 1,200 years of the demigods.[74]

4,800 + 3,600 + 2,400 + 1,200 equals the 12,000 years of the demigods mentioned above. 12,000 × 360 equals 4,320,000 earth solar years for one *divya-yuga* (*yuga* cycle). A thousand *yuga* cycles makes one day of Brahmā, or 4,320,000,000 years. Note that in his translation of this particular verse, Prabhupāda uses the word "millennium" to refer to the individual "ages" that

67 Prabhupāda, *Caitanya-caritāmṛta, Madhya-līlā* 20.246, purport, 1498.
68 Prabhupāda, *Śrīmad-Bhāgavatam*, 1.3.15, purport, 29.
69 Prabhupāda, *Śrīmad-Bhāgavatam* 3.11.25, 520.
70 Prabhupāda, *Śrīmad-Bhāgavatam* 3.11.25, purport, 520.
71 Prabhupāda, *Śrīmad-Bhāgavatam* 3.11.12, 517.
72 Prabhupāda, *Śrīmad-Bhāgavatam* 3.11.19, purport, 519.
73 Prabhupāda, *Śrīmad-Bhāgavatam* 3.11.18, 518-519: "Maitreya said: O Vidura, the four millenniums are called the Satya, Tretā, Dvāpara and Kali yugas. The aggregate number of years of all of these combined is equal to twelve thousand years of the demigods."
74 Prabhupāda, *Śrīmad-Bhāgavatam* 3.11.19, 519.

comprise the *divya-yuga*, rather than the *divya-yuga* itself. As some of these Sanskrit terms have no equivalent in English, conceptually or otherwise, Prabhupāda makes use of English words that have generally similar meanings to the Sanskrit terminology, such as when he uses the word "universe" for *brahmāṇḍa*. The purport to the above verse is as follows:

> As aforementioned, one year of the demigods is equal to 360 years of the human beings. The duration of the Satya-yuga is therefore 4,800 × 360, or 1,728,000 years. The duration of the Tretā-yuga is 3,600 × 360, or 1,296,000 years. The duration of the Dvāpara-yuga is 2,400 × 360, or 864,000 years. And the last, the Kali-yuga, is 1,200 × 360, or 432,000 years.[75]

It is from these figures that the lifespan of a Manu can be reckoned. As a Manu lives for 71 *divya-yugas*, "Each Manu lives 4,320,000 years multiplied by 71."[76] "This is about 852,000 years in the calculation of the demigods," (12,000 demigod years × 71) "or, in the calculation of human beings, 306,720,000 years."[77] The present Manu has already lived for 4,320,000 years multiplied by 28,"[78] which equals 120,960,000 earth years.

The above numbers, however, conflict with what Prabhupāda wrote elsewhere, such as in *Bhagavad-gītā As It Is*. There he states, "the age of the current Manu (Vaivasvata) is calculated to last some 305,300,000 years, of which 120,400,000 have passed,"[79] these figures being 1,420,000 and 560,000 years less than those given above. In the same purport to verse 1 of the 4th chapter of *Bhagavad-gītā*, the figures given for the lengths of the four *yugas* are also rounded down, so it appears that these numbers are more approximations. Here, although he cites 432,000 years as the duration of Kali-yuga, he gives 800,000 and 1,200,000 years as the lengths of Dvāpara-yuga and Tretā-yuga respectively, instead of 864,000 and 1,296,000 years. The figure of 120,400,000 years having elapsed in the current *manvantara* is also given in Prabhupāda's purport to Canto 6, Chapter 10, Text 16 of the *Śrīmad-Bhāgavatam*, a verse mentioned earlier.[80]

The explanation for this apparent discrepancy appears to be that the lower figures are arrived at by using an alternate number for the duration of a *divya-yuga* of 4,300,000 instead of 4,320,000, while the higher figures do

75 Prabhupāda, *Śrīmad-Bhāgavatam* 3.11.19 purport, 519.
76 Prabhupāda, *Śrīmad-Bhāgavatam* 4.30.49 purport, 1189.
77 Prabhupāda, *Śrīmad-Bhāgavatam* 3.11.24 purport, 520.
78 Prabhupāda, *Śrīmad-Bhāgavatam* 4.30.49 purport, 1189.
79 Prabhupāda, *Bhagavad-gītā* 4.1, 215.
80 Prabhupāda, *Śrīmad-Bhāgavatam* 6.10.16 purport, 1539.

use 4,320,000. 4,300,000 appears to be a rounded down approximation or rough estimate of the number of earth years in a *divya-yuga* which, if multiplied by 1000 (for the 1000 *divya-yugas* in a day of Brahmā), we would get an approximate length for the *kalpa* (day of Brahmā) of 4,300,000,000 years, as used by Prabhupāda, for example, in a 1966 lecture to his early disciples in New York.[81] Indeed, if we divide the figure of 305,300,000 (alternate duration of a *manvantara*) by 4,300,000 (alternate duration of a *divya-yuga*) we get 71, and if we divide 120,400,000 (alternate number of years passed in *Vaivasvata-manvantara*) by 4,300,000 we get 28. When Prabhupāda was discussing this topic in general terms, then he used approximate figures rounded to the nearest 100,000 (one hundred thousand = *lakh*—a commonly used numerical term in Indian culture). Other times he was more specific and used the exact figures. An example of Prabhupāda using the approximate values in combination with the words "about" and "*lakh*" is found in his lectures:

> So, each *yuga* is divided into four. Satya-yuga is about eighteen *lakhs*. Eighteen *lakhs* means 1,800,000's of years. That is Satya-yuga, 1,800,000's. And the next *yuga*, Tretā, is 1,200,000's of years; and the next *yuga*, Dvāpara, 800,000's of years; and this Kali-yuga is 400,000's of years ...[82]

> Satya-yuga, eighteen *lakhs*; then Tretā-yuga, twelve *lakhs*; then Dvāpara-yuga, eight *lakhs*; and Kali-yuga, four *lakhs*. Altogether forty-three.[83]

Another way of calculating the duration of a *manvantara* is to divide 1000 *divya-yugas* into 14 parts. 1000 divided by 14 = 71.4 *divya-yugas*. Equivalent to 0.4 of a *divya-yuga* is one *Satya-yuga*, or 1,728,000 Earth years. In a hand-written 1957 essay on the *Caitanya-caritāmṛta*, Prabhupāda stated: "One day of Brahmā is equal to ... fourteen Manus plus 15 times the age of a Satya yuga. This duration of one day's time in the life of Brahmā, is called a Kalpa."[84]

81 A. C. Bhaktivedanta Swami Prabhupāda, '661223 - Lecture CC Madhya 20.318–329 - New York', *Vanisource* (2020) <https://vanisource.org/wiki/661223_-_Lecture_CC_Madhya_20.318–329_-_New_York> [accessed August 15, 2023].

82 A. C. Bhaktivedanta Swami Prabhupāda, '661223 - Lecture CC Madhya 20.318–329 - New York', *Vanisource* (2020), <https://vanisource.org/wiki/661223_-_Lecture_CC_Madhya_20.318-329_-_New_York> [accessed August 15, 2023].

83 A. C. Bhaktivedanta Swami Prabhupāda, '751029 - Lecture SB 03.28.19 - Nairobi', *Vanisource* (2021), <https://vanisource.org/wiki/751029_-_Lecture_SB_03.28.19_-_Nairobi> [accessed August 15, 2023].

84 A. C. Bhaktivedanta Swami Prabhupāda, '*Caitanya-caritāmṛta*, Adi-līlā, Chapter 3 (Handwritten)', *Vanisource* (2016), <https://vanisource.org/wiki/Caitanya-caritamrta,_Adi-lila,_Chapter_3_(Handwritten)> [accessed August 15, 2023].

This refers to *sandhyās*, transitional periods between each *manvantara*, one at the beginning and one at the end of the sequence, and thirteen between the fourteen *manvantaras*, making fifteen altogether; Prabhupāda indicates that each lasts for the same length of time as a Satya-yuga, or 0.4 of a *divya-yuga*. Figures 3 and 4 incorporate this detail. If one *sandhyā* is assigned to each *manvantara*, then the length of each *manvantara* is 71.4 *divya-yugas*. This agrees with Thompson's explanation,[85] wherein he refers to the use of the phrase "a little more than" in the following verse in *Śrīmad-Bhāgavatam*:

> Each and every Manu enjoys a life of a little more than seventy-one sets of four millenniums.[86]

Figure 4: Diagram Illustrating a *Manvantara*.

A *manvantara* consists of 71 *divya-yugas*. We are currently in the 28th *divya-yuga* of the reign of Vaivasvata Manu, the seventh *manvantara* in the current day of Brahmā.

The blue segment in the diagram in Figure 4 represents the *sandhyā*, or junction, that occurs between adjacent *manvantaras*.

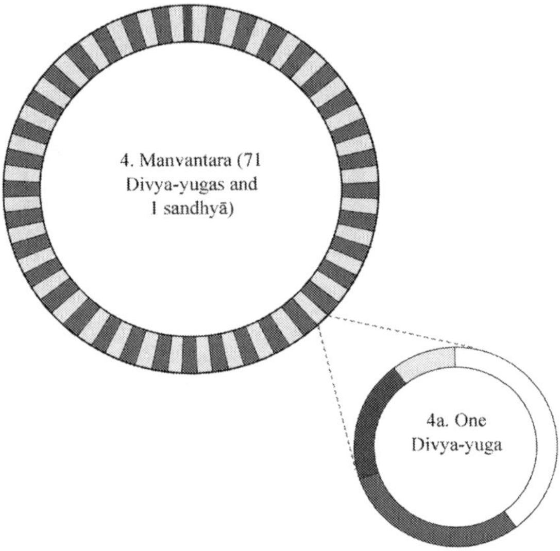

85 Richard L. Thompson, *Mysteries of the Sacred Universe: The Cosmology of the Bhagavata Purana*, (Alachua, Florida: Institute for Vaishnava Studies, 2016), 227.

86 Prabhupāda, *Śrīmad-Bhāgavatam* 3.11.24, 519.

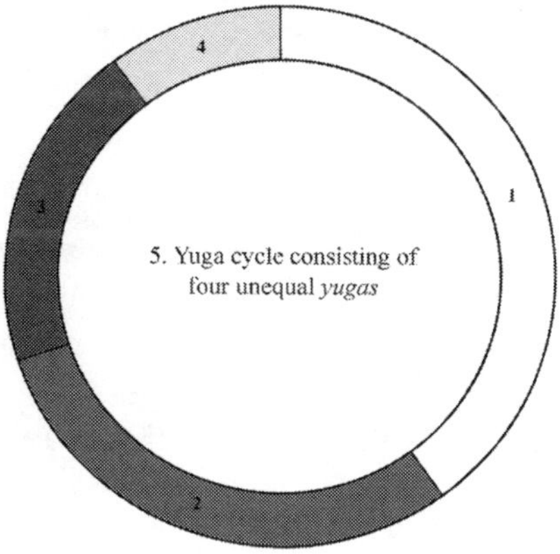

Figure 5: Diagram Illustrating a *Yuga* Cycle (*Divya-yuga*).

Key to the diagram in Figure 5: 1. Satya-yuga 2. Tretā-yuga 3. Dvāpara-yuga 4. Kali-yuga.

The diagram in Figure 5 represents the current *yuga* cycle, which is the 28th *divya-yuga* of the reign of Vaivasvata Manu. In this particular *divya-yuga*, the third age, Tretā-yuga, appears before the second (Dvāpara). It is when this overlapping occurs that Lord Kṛṣṇa appears in His original form. Following Lord Kṛṣṇa's appearance, in the same *yuga* cycle, Lord Caitanya appears in the Kali-yuga. Both Lord Kṛṣṇa and Lord Caitanya appear only once in each day of Brahmā (*kalpa*) in this specific *yuga* cycle (28th of Vaivasvata).

Not shown on the diagram are the *sandhyās*, or junctions, that occur between each *divya-yuga* and between each *yuga*. At the end of each *divya-yuga* a devastation takes place.

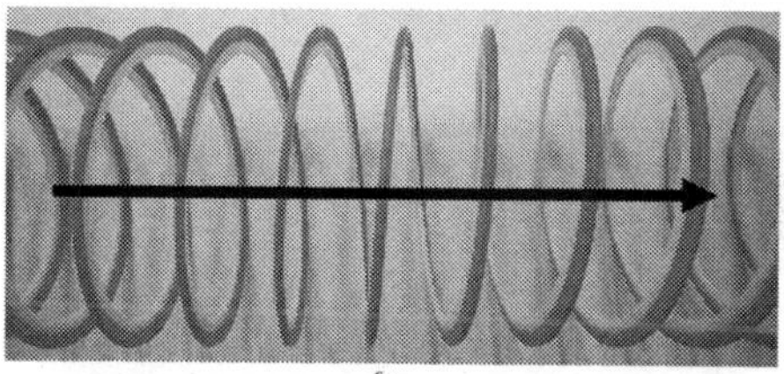
6a

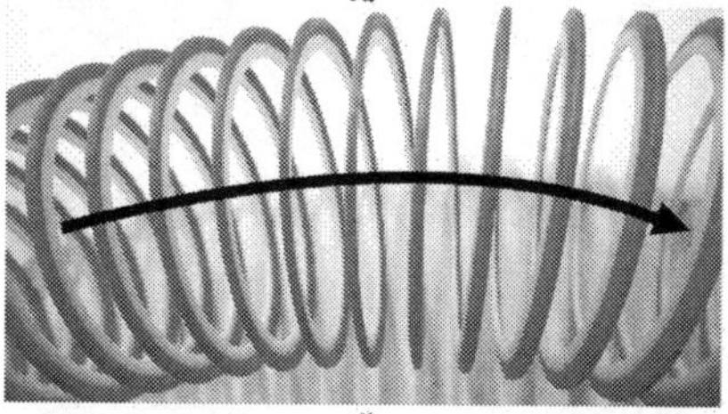
6b

6c

Figure 6: Forward Motion in Cyclical Time.

6a. Upon completion of each cycle of time, there has also been forward movement. Although time in the material world in the Vedic system is described as cyclical, there is also linear movement as the material cosmic manifestation, under the influence of time, moves from its coming into being through to its end.

6b. The Vedic system describes time in terms of cycles within cycles at every level. Therefore, in this model, the forward linear movement of time is actually curvilinear.

6c. The 'coil' or spiral created by one cycle is itself in a larger coil created by the next higher level of time. If we consider modern astronomical knowledge and measurement of time, with its basis in the movement of the moon around the earth and the earth around the sun, these movements can also be

thought of as spirals through space-time, as the sun moves in its orbit around the galactic hub as the galaxy moves through the universe.

Summary of calculations in the teachings of A. C. Bhaktivedanta Swami Prabhupāda

Satya-, Dvāpara-, Tretā-, and Kali-yugas added together (1 *divya-yuga*) total 4,320,000 solar years. 1,000 *divya-yugas* make 1 day (12 hours) of Brahmā (*kalpa*) = 4,320,000,000 solar years. Twice this amount (to account for Brahmā's night) makes 24 hours of Brahmā 8,640,000,000 solar years. Brahmā's year has 360 days so 8,640,000,000 multiplied by 360 = 3,110,400,000,000 solar years. Brahmā lives for 100 of his years, which is 3,110,400,000,000 multiplied by 100, or 311,040,000,000,000 solar years, and this is the duration of the life of the universe.

Table 3. Summary of time equivalents.

Earth Time (Solar Years)	Higher Planets (Demigod Time)	Manus	Lord Brahmā
1	1 day & night		
360	1 year of the demigods		
4,320,000	1 *divya-yuga* (12,000 years of the demigods)		
306,720,000	71 *divya-yugas* (852,000 years of the demigods)	1 *manvantara*	
4,320,000,000	1,000 *divya-yugas* (12,000,000 years of the demigods)*	14 *manvantaras*	1 *kalpa* (1 twelve-hour day of Brahmā)
8,640,000,000			24 hours of Brahmā (1 day and 1 night)
259,200,000,000		420 *manvantaras*	30 *kalpas* (1 month of Brahmā)

Earth Time (Solar Years)	Higher Planets (Demigod Time)	Manus	Lord Brahmā
3,110,400,000,000		5040 *manvantaras*	360 *kalpas* (12 months of Brahmā) (1 year of Brahmā)
			50 years of Brahmā (1 *parārdha*[87])
311,040,000,000,000		504,000 *manvantaras*	100 years of Brahmā (1 *vikalpa*, duration of the universe)

*100 years on Maharloka

Table 4. Details pertaining to the four *yugas* in a *divya-yuga*.

Age	Demigod Years	Solar Years	Color of Yuga-avatāra[88]
Satya-yuga	4,800	1,728,000	*śukla* (white) [Hayagriva]
Tretā-yuga	3,600	1,296,000	*rakta* (red)
Dvapara-yuga	2,400	864,000	*śyāma* (dark blue, blue-black, or black[89])
Kali-yuga	1,200	432,000	generally, *kṛṣṇa* (black) but in special cases *pīta* (yellow[90])

Standard Time[91]

Chapter 11 of the Third Canto of *Śrīmad-Bhāgavatam* is entitled "Calculation of Time, from the Atom." Prabhupāda uses the word "atom" as a translation for a Sanskrit term, *anu*, which means something like "the smallest particle,"but "atom" is a word familiar to Westerners and which has a similar meaning, while not necessarily equating exactly to what the word "atom" means to a Western scientist. Prabhupāda explains that it "is almost the same as the modern science of atomism." Interestingly, the first verse of Chapter 11 of

87 Prabhupāda, *Śrīmad-Bhāgavatam* 3.11.35 purport, 522.
88 Prabhupāda, *Caitanya-caritāmṛta, Madhya-līlā* 20.330, 1515.
89 Prabhupāda, *Caitanya-caritāmṛta, Madhya-līlā* 6.101, 819.
90 Prabhupāda, *Caitanya-caritāmṛta, Madhya-līlā* 6.103, 819.
91 The quotations and information in this section up through the list of the time units comes from Prabhupāda, *Śrīmad-Bhāgavatam* 3.11.1–6, 515–16, unless otherwise referenced.

the Third Canto states that although invisible, the atom always exists, "even after the dissolution of all forms," with Prabhupāda adding, "The atom is the minute subtle form of eternal time ... The material manifestation's ultimate particle, which is indivisible and not formed into a body, is called the atom." The material body is described as "a combination of such atoms." Indeed, verse 2 states:

> Atoms are the ultimate state of the manifest universe. When they stay in their own forms without forming different bodies, they are called the unlimited oneness. There are certainly different bodies in physical forms, but the atoms themselves form the complete manifestation.

Śrīmad-Bhāgavatam gives a system for estimating time based initially on the time taken for the sun to traverse an atom or a particular combination of atomic particles. Prabhupāda remarks that, "The sun is the central point for calculating all different durations," time and space being correlated terms.

Overall, "That time which covers the unmanifest aggregate of atoms is called the great time ... The complete calculation of the time of creation, maintenance and dissolution, measured in terms of the circulation of the total planetary systems until the end of creation, is known as the supreme *kāla*,"[92] with all subdivisions of time calculable "according to the movements of the sun (*savitā*)."[93]

Following is a list of the time units given in *Śrīmad-Bhāgavatam*:[94]

- » "The atom is an invisible particle." Two atoms make one double atom, and three double atoms make one hexatom, or *trasareṇu*. The time taken "for the integration of three *trasareṇus* is called a *truṭi*."
- » "One hundred *truṭis* make one *vedha*."
- » "Three *vedhas* make one *lava*."
- » "Three *lavas* is equal to one *nimeṣa*."
- » "... three *nimeṣas* makes one *kṣaṇa*.
- » "Five *kṣaṇas* ... make one *kāṣṭhā*."
- » "Fifteen *kāṣṭhās* make one *laghu*."
- » "Fifteen *laghus* make one *nāḍikā*" (also known as a *daṇḍa*).
- » "Two *daṇḍas* make one *muhūrta*."

92 Prabhupāda, *Śrīmad-Bhāgavatam* 3.11.14, verse and purport, 518.

93 A. C. Bhaktivedanta Swami Prabhupāda, *Śrīmad-Bhāgavatam: Tenth Canto-Part One*, (Los Angeles, CA: Bhaktivedanta Book Trust, 1977), 10.3.26 purport, 238.

94 The values in the following list were derived from Prabhupāda, *Śrīmad-Bhāgavatam* 3.11.5–12, 516–17.

- "Six or seven *daṇḍas* make one fourth" (*prahara* also called *yāma*) of a human "day or night."
- "Fifteen days and nights are a" *pakṣa* (fortnight).
- There are two *pakṣas*, "white and black, in a month." One month in human terms is "one complete day and night for the Pitā planets."
- A season is "two of such months."
- "Six months comprise one complete movement of the sun from south to north."
- "Two solar movements make one day and night of the demigods. and that combination of day and night is one complete calendar year" of twelve months (one full orbit of the sun) for the human being, who is expected to live up to one hundred years.

Śrīla Prabhupāda states that, by calculation, one *laghu* equals two minutes. "Atomic" Vedic time calculations may be converted into current Western time unit usage with this understanding.[95] In this instance, Prabhupāda uses the terminology associated with Western time units to describe, in approximate and familiar terms, the Vedic time divisions rather than give their actual time equivalents, i.e. if a *muhūrta* is called an hour then the other Vedic time units would have the corresponding values given in his purport to *Śrīmad-Bhāgavatam* 3.11.14. Elsewhere he refers to a *muhūrta* as being equal to 48 minutes, which gives a different set of durations for the Vedic time units by deduction from this fact (See Table 5).

In *Arcana-paddhati*, a manual on deity worship produced by disciples of Prabhupāda in 1978, it is stated:

> ... twenty-four minutes is equal to one *daṇḍa*. Two *daṇḍas*, or forty-eight minutes, is equal to one *muhūrta*. In the day and the night together, there are a total of thirty *muhūrtas*. In the last portion of the night, the time beginning two *muhūrtas* before the rise of the sun up to the rise of the sun, or one hour and thirty-six minutes before the rise of the sun, is called *aruṇodaya*. Of these two *muhūrtas*, the first *muhurta* is called the *brahma-muhūrta*.[96]

This statement reflects references in Prabhupāda's books given in this paper for details in the right-hand column of Table 5.

95 Prabhupāda, *Śrīmad-Bhāgavatam* 3.11.7, purport, 516.

96 Jayaśacīnandana dāsa Adhikārī and Jayatīrtha dāsa Adhikārī, *Arcana-paddhati* (Los Angeles, CA: Bhaktivedanta Book Trust, 1978), 1.

Table 5. Summary of time divisions given in *Śrīmad-Bhāgavatam*.

Unit	Duration (Taking 1 *muhūrta* as 1 hour)[97, 98]	Duration (Taking 1 *muhūrta* as 48 minutes)[99, 100, 101, 102]
One *truṭi*	0.000593 seconds	0.000474 seconds
One *vedha*	0.059 seconds	0.0474 seconds
Ona *lava*	0.178 seconds	0.142 seconds
One *nimeṣa*	0.53 seconds	0.427 seconds
One *kṣaṇa*	1.6 seconds	1.28 seconds
One *kāṣṭha*	8 seconds	6.4 seconds
One *laghu*	2 minutes	1 minute 36 seconds
One *daṇḍa* (*nāḍikā*)	30 minutes	24 minutes
One *muhūrta*	1 hour	48 minutes
One *prahara*	3 hours	
One day or night	12 hours	15 *muhurtas*
One full day	24 hours	30 *muhurtas*
One *pakṣa*	15 days	

Śrīmad-Bhāgavatam provides a description of a practical method for measuring a standard unit of Vedic time, in this case a *daṇḍa*:

> "The measuring pot for one *nāḍikā*, or *daṇḍa*, can be prepared with a six-*pala*-weight [fourteen ounce] pot of copper, in which a hole is bored with a gold probe weighing four *māṣa* and measuring four fingers long."
>
> "This regulates the diameter of the hole."

97 Prabhupāda, *Śrīmad-Bhāgavatam* 3.11.14 purport, 518.

98 A. C. Bhaktivedanta Swami Prabhupāda, '701214 - Conversation A - Indore', *Vanisource* (2023): "*Brāhma-muhūrta*, this hour, one hour before sunrise, is very auspicious moment." <https://vanisource.org/wiki/701214_-_Conversation_A_-_Indore> [accessed August 15, 2023].

99 Prabhupāda, *Śrīmad-Bhāgavatam* 3.20.46 purport, 615: "The time early in the morning, one and a half hours before sunrise, is called *brāhma-muhūrta*."

100 Prabhupāda, *Śrīmad-Bhāgavatam* 5.21.12 word-for-word translation, 1386: "*muhūrtena*—in a *muhūrta* (forty-eight minutes)"

101 Prabhupāda, *Śrīmad-Bhāgavatam* 5.24.3 purport, 1397: "*Rāhu* cannot stay in front of the sun or moon for more than a *muhūrta* (forty-eight minutes)."

102 Prabhupāda, *KRSNA*, Ch. 33 "Description of the Rāsa Dance," 278: "The *brāhma-muhūrta* takes place about one and a half hours before sunrise."

> "When the pot is placed on water, the time before the water overflows in the pot is called one *daṇḍa*."[103]

Prabhupāda explains that this technique for measuring time is similar to the use of sand in an hourglass, and that in Vedic civilization the simplest methods for calculating time were utilized (see footnote 101).

The Sun-god's Chariot Wheel

Śrīmad-Bhāgavatam describes the solar year using the metaphor of the sun-god's chariot wheel. Verse 13 in Chapter 21 of the Fifth Canto begins, "The chariot of the sun-god has only one wheel, which is known as Saṁvatsara."[104] Prabhupāda elsewhere gives the meaning of the word *saṁvatsara* as "the progress of time."[105] In the case of the sun, applied to this verse, it would refer to the year-long movement of the sun in relation to the cosmic mountain, Meru, and in relation to the earth against the background of stars from a geocentric perspective, a 360-day year in Vedic culture. The Fifth Canto verse continues with a depiction of the parts of the year as components of a wheel: "The twelve months are calculated to be its twelve spokes, the six seasons" (*ṛtus*) "are the sections of its rim ...," which, in a Third Canto verse, is described as having "numberless leaves" (moments) "carved upon it."[106] "The three *cātur-māsya* periods" (periods of four months) "are its three-sectioned hub" (see Figure 7).[107]

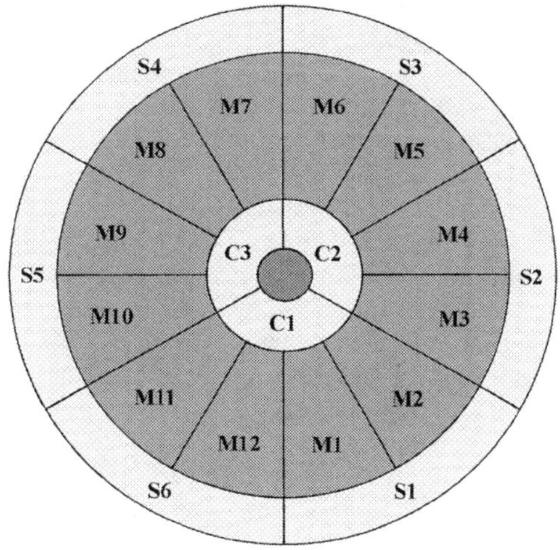

Figure 7: Diagram Illustrating the Sun-god's Chariot Wheel.
C1-3: The three *cātur-māsya* (four-month) periods in the solar year are represented by the three-sectioned hub of the chariot wheel. C3 is the *cātur-māsya* period ending in Kārttika, during which specific fasts are observed by Vaiṣṇavas: Śrāvaṇa (M7)—green

103 Prabhupāda, *Śrīmad-Bhāgavatam* 3.11.9 text and purport, 517.

104 Prabhupāda, *Śrīmad-Bhāgavatam* 5.21.13, 1387.

105 Prabhupāda, *Śrīmad-Bhāgavatam* 4.29.18–20 purport, 1148.

106 Prabhupāda, *Śrīmad-Bhāgavatam* 3.21.18, 621.

107 Prabhupāda, *Śrīmad-Bhāgavatam* 5.21.13, 1387.

leafy vegetables; Bhādra (M8)—yogurt; Āśvin (M9)—milk; and Kārttika (M10)—urad dahl.

M1-12: The twelve months of the 360-day Vedic solar year, and S1-6: The six seasons of the 360-day Vedic solar year.

M1	Magha (Jan/Feb) [Mādhava]	S1	Śita (winter)
M2	Phālguna (Feb/Mar) [Govinda]		
M3	Caitra (Mar/Apr) [Viṣṇu]	S2	Vāsanta (spring)
M4	Vaiśakha (Apr/May) [Madhusūdana]		
M5	Jyaiṣṭha (May/Jun) [Trivikrama]	S3	Uṣṇa or Gṛṣma (summer)
M6	Āṣāḍha (Jun/Jul) [Vāmana]		
M7	Śrāvaṇa (Jul/Aug) [Śrīdhara]	S4	Varṣā (rainy season)
M8	Bhādra (Aug/Sep) [Hṛṣīkeśa]		
M9	Āśvin (Sep/Oct) [Padmanābha]	S5	Śārada (autumn, fall)
M10	Kārttika (Oct/Nov) [Dāmodara]		
M11	Margaśirṣa (Nov/Dec) [Keśava]	S6	Hemant (pre-winter season)
M12	Pauṣa (Dec/Jan) [Nārāyaṇa]		

The predominating Deities of each month are included in the table next to the names of Their respective months. These twelve predominating Deities are the expansions of the *catur-vyūha*. The *catur-vyūha* are Vāsudeva, Saṅkarṣaṇa, Pradyumna, and Aniruddha. There are three expansions of each catur-vyūha form, named differently according to the position of the weapons that are held in the four hands of the Deities. The Vāsudeva expansions are Keśava, Nārāyaṇa, and Mādhava. The expansions of Saṅkarṣaṇa are Govinda, Viṣṇu and Madhusūdana (this Govinda is not the original Govinda, son of Mahārāja Nanda). The expansions of Aniruddha are Hṛṣīkeśa, Padmanābha, and Dāmodara. These same twelve Viṣṇu expansions are those whose names are recited when applying *tilaka* to the body. The names of these predominating Deities are used by the International Society for Krishna Consciousness as the names of the months on the Vaiṣṇava calendar. Therefore, for example, Māgha-māsa is called Mādhava-māsa, etc.

The wheel constructed from the details given in Canto 5, Chapter 21, Text 13 of *Śrīmad-Bhāgavatam* shown in Figure 7 has twelve spokes, but in the Third Canto, Chapter 21 Text 18, which also gives a description of this wheel, it says, "It has thirteen spokes, 360 joints ..." The 360 joints clearly refer to the 360-day year of the Vedic calendar. Regarding the thirteen spokes, Prabhupāda explains in the purport to this verse:

> There are six seasons, called *ṛtus*, and there is the period of four months called *cāturmāsya*. Three periods of four months complete one year. According to Vedic astronomical calculations, there are thirteen months. The thirteenth month is called *adhi-māsa* or *mala-māsa* and is added every third year.[108]

The thirteenth month is therefore one that is added every third year, making a leap year that brings the calendar back into line with the solar year. That this wheel is intended as a metaphor is supported by the analogy that completes the Fifth Canto verse:

> One side of the axle carrying the wheel rests upon the summit of Mount Sumeru, and the other rests upon Mānasottara Mountain. Affixed to the outer end of the axle, the wheel continuously rotates on Mānasottara Mountain like the wheel of an oil-pressing machine.[109]

Mount Meru (or Sumeru) and Mānasottara Mountain are features of the earthy plane of existence (Bhū-maṇḍala or "earth-disk"[110]) in the *brahmāṇḍa*, Meru being at the center, and Mānasottara Mountain being a ring-shaped structure on one of the concentric, ring-shaped islands of Bhū-maṇḍala that centers on Meru. The Sun-god's chariot wheel is also said to rotate "on the eternal existence, Brahman."[111] The last sentence of the translation of the Fifth Canto verse above uses a reference to an oil-pressing machine to illustrate the idea that one revolution of the Sun-god's chariot wheel around Mānasottara Mountain takes one solar year, represented by the circumference of the wheel, and covering the full circular length of Mānasottara Mountain, which from the perspective of our earth, takes the sun through the constellations of the zodiac. This could suggest that the circumference of the wheel is equal to the circumference of the circular Mānasottara Mountain on Bhū-maṇḍala. The next verse states:

> As in an oil-pressing machine, this first axle is attached to a second axle, which is one-fourth as long [3,937,500 yojanas, or 31,500,000 miles]. The upper end of this second axle is attached to Dhruvaloka by a rope of wind.[112]

108 Prabhupāda, *Śrīmad-Bhāgavatam* 3.21.18 purport, 621.
109 Prabhupāda, *Śrīmad-Bhāgavatam* 5.21.13, 1387.
110 Thompson, *Mysteries of the Sacred Universe*, 135.
111 Prabhupāda, *Śrīmad-Bhāgavatam* 3.21.18 purport, 621.
112 Prabhupāda, *Śrīmad-Bhāgavatam* 5.21.14, 1387.

The "first axle" referred to is the one mentioned in the previous verse. The second axle is the one that is visualized as pointing up from the top of Mount Meru towards the pole star (Dhruvaloka), and this axle rotates as the fixed first axle, around which the wheel rotates, moves. This description using the metaphor of the oil-pressing machine emphasizes the cyclical nature of time and events in the material world.

This reference to an oil pressing machine possibly describes the type of stone oil press used for millennia in the Mediterranean and Middle East regions for crushing olives prior to pressing out the olive oil (Figure 8). However, traditional oil presses in use in India seem to be of the *ghani* or *kal sekku* or *kal chekku* type, which are like a large mortar and pestle for extracting oil from products such as mustard seeds, and which may be operated by people or by animals such as oxen. The type of oil press with a large stone wheel has, however, been observed in rural India by Jayādvaita Swami (personal communication). Nevertheless, further research is needed to determine exactly which type of oil press the Fifth Canto refers to.

Figure 8: Middle Eastern Stone Oil Press for Crushing Olives.

The stone oil press in Figure 8 is possibly something like the kind described in the Fifth Canto of *Śrīmad-Bhāgavatam*, with two axles, one pointing upwards and the other at right angles to it. The axle pointing upwards is said in the *Bhāgavatam* to be one fourth the length of the other, and such a version of the device depicted does exist in the Mediterranean area.

Summary

The Vedic time system as described in the texts translated by A. C. Bhaktivedanta Swami Prabhupāda, and in his commentaries on those texts, can be divided into two components. There are divisions of time that are literally on a cosmic scale, and for which there would be no observable measures for practical purposes, although they can be calculated in terms of time divisions that are observable. That is the second group of divisions of time, those that are measurable with observable events, such as the apparent movement of the sun in the sky. Smaller time units in the Vedic system, as are the largest, are calculated with reference to the sun, and even the smallest, which would be too brief to notice with unaided human minds and senses, are nevertheless related to the movement of the sun over minute particles, and ultimately the smallest particle of all, the *anu*, translated by Prabhupāda as the "atom."

Time, *kāla*, is seen by the Vedic texts as one of the expansions of the Supreme Personality and is equated with the omnipresent form of Viṣṇu known as Kṣīrodakaśāyī Viṣṇu, the Supersoul. In combination with space and consciousness, *kāla* makes things happen in the material cosmos. Vedic texts generally envisage forces and energies at work in the cosmos in terms of personalities, either versions of the Supreme Personality or parts and parcels of that Personality, each playing out their roles in the grand cosmic drama of the material world. The scale of events in this drama equates favorably with current understandings of modern science, despite the seemingly disparate world views of the *Vedas* and Western science.

Phenomenology as a Cross-cultural Bridge Between Science and Religion, East and West

Alexey Timoschuk
Vladimir Institute of Law

Presented at the
Consciousness in Science Conference
January 18–20, 2019

"East is East and West is West, and never the twain shall meet" – that is the classic poetic formula by Rudyard Kipling of the fundamental opposition of cultures, interpreted as a conflict of Eastern and Western worldviews. One indologist called traditional texts "mythology that we can only try to reconstruct ... jumble of paradoxes heaped on paradoxes, tropes heaped on tropes."[1] We consider E. Husserl's phenomenology as standing apart from other Western philosophies, because it can be applied to heterogeneous cultural experience by bracketing our accumulated stereotypes.

Phenomenology is also a promising tool due to its compatibility with Eastern psychotechnics since it was started with the idea that reality of the mind is apodictic – it is a first given reality. Yoga, Vedānta, and Buddhism offer prolific material on the study of consciousness. Nowhere in the world during the pre-scientific period do we find such a rich terminology, containing the finest nuances as that which refers to *ātmā,* or the internal self.

We will begin by examining concepts of *self* found in Vedānta, a religious and philosophical school in India, concentrating on the issue of the *mind.* Three states of consciousness (*jāgrata* – awakeness, *svapna* – sleep, *suṣupti* – deep sleep) are considered to be simply different manifestations of the single state, sleep. Even while awake one is dreaming, being plunged into the reveries of the subconscious mind. The Vedānta school considers all these phases to be manifestations of the implicit, non-deductive fourth state of

1 The Ṛg Veda, "One hundred and eight hymns," O Flaherty, Wendy Doniger (editor), London (1981) p. 18.

mind – *turīya* (*Bhāgavata Purāṇa* 11.25.20). This stage of self-reflection, said to be a conscious experience of the universal self, *turīya* is the desirable goal of all mystic and reflective practices in India. Self-realization is defined in Vedānta as a balance between the practical and theoretical side of introspection within the mind.

We shall examine two schools of Vedānta – Advaita and Dvaita. The first entertains the ontology of nonqualified monism, and the second – qualitative pluralism. Advaita Vedānta considers that the external world, consisting of objects and perceived by the person in everyday experience, does not exist by itself and is merely a product of the mind projecting outside itself. The person as an individual, separate personality is not real. The real essence of human existence consists in its unqualified oneness with God.

Dvaita Vedānta, on the other hand, concedes the illusory nature of this world only in connection with its temporary nature. Illusion is the acceptance of the transient as the eternal, as all in all. There is a well-known example concerning Advaita Vedānta's affirmation of the non-existence of the external world: a man mistaking a rope for a snake. Dvaita Vedānta explains this example by pointing out that all three – the man, the rope, and the snake – do in fact exist, even though the man mistakenly thinks that the rope is a snake. The individuality of man, according to Dvaita Vedānta, is an eternal reality just as is the individuality of God: "Never was there a time when I did not exist, nor you, nor all these kings; nor in the future shall any of us cease to be ." (Bhagavad-gītā 2.12)

All schools of Vedānta practice mind and speech control, which can be compared with abstinence from judgments (epoché) in phenomenology. Husserl borrowed the term from the philosophy of skeptics. Diogenes Laertius characterizes the practice of the epoché as follows:

> The ultimate goal of skeptics is to abstain from judgments (*epoché*), beyond which, like a shadow, tranquility (*ataraxy*) follows. That's what the followers of Timon and Enesidema say. In fact, we prefer to avoid only those things that depend on us, and not what comes inevitably, like hunger, thirst, and pain, that we can not avoid because our judgements are useless here. Dogmatists claim that the skeptics in their lifestyles will not refrain from devouring their own father, if so required; but skeptics reply that they follow their way of life and that they shall refrain from dogmatic questions, but not from those that are everyday and ordinary. Therefore, in the latter, one can prefer and avoid something, following the customs and respecting

> the laws. However, others say that the ultimate goal for the skeptics is dispassion, and others – it's softness.[2]

Phenomenological reduction is the operation of removing the attributes of objects from the acts of consciousness. It consists of three steps – psychic, eidetic, and transcendental reductions. Psychic reduction is the elimination of assessments, motivations, and intentions related to the subject. Eidetic reduction removes the form of cognitive operations, and with transcendental reduction one focuses his mind on its own ground. *Epoché* is the important method of phenomenological reduction, since during the ascent to the unconditional state of consciousness, you must learn to hold back the quality of the object removed from its intentional attraction.

The Vedānta school practices several levels of *epoché*. First: to abstain from ordinary statements. The practice of abstinence from speech (*mauna*) is widely practiced in both Vedānta schools. In Advaita Vedānta, complete abstinence from verbal utterances is practiced. There are cases when *mauna* is observed for many years. In 1971 at the festival of Kumbha-melā, which is held every twelve years in India, a *mauni* who practiced the vow of silence for 20 years is said to have been present. Thus, Advaita will not agree with E. Husserl that intentionality is the attribute of mind, since its goal is a trans-cognitive state of non-thinking, where the object, subject, and the process are the same. Non-dual existence of individualities is the basic credo of this school.

In Dvaita Vedānta, *mauna* is a specific instantiation, namely, to speak only about Kṛṣṇa or spiritual topics, and to abandon everyday judgments. The next level of the *epoché* – refraining from mental speculation or sophistry (*jalpa*) – is practiced only in Dvaita Vedānta, where the Personality of Godhead is recognized and the practitioner places his faith in such representatives of God as the Holy Scriptures (*Bhagavad-gītā, Bhāgavata Purāṇa*), and Guru-Vaiṣṇava (devotee of God). Advaita Vedānta, by contrast, pays great attention to grammar and sophistry, plus dialectic and rhetoric, to maintain the transcendental level of consciousness (*paramārthika*).

Both schools of Vedānta recognize transmigration of the soul and agree that the mind carries from one life to the next the imprints of sensual images (*saṁskāras*), which can be both virtuous and sinful. *Saṁskāras*, which are like anchors, arise from direct contact with the objects of the senses and include reason, motive, media, and external communications. It doesn't matter which anchors are left in the mind of man, either virtuous or sinful, his restless subtle body leads the soul to incarnate again. Therefore, the subtle body in Vedānta is also called *kāraṇa deha*, causal body.

2 Diogenes Laertius, On the Life, Teachings and Sayings of Famous Philosophers, Book 9.

A similar description of consciousness is found in Yoga. Patañjali in the *Yoga-sūtras* describes the human mind (*citta*) as a very restless, being constantly in certain affective states (*kleśas*). These states are retained in the minds of individual as imprints (*saṁskāras*). If the number of similar *saṁskāras* accumulates, they evolve to persistent desire (*vāsanā*), and these eventually manifest as behavior (*vṛtti*). Thus is formed the wheel of conditioned thinking and acting. *Citta-vṛtti-nirodha* – this is a recipe for Yoga, which is to unlock the social and individual automatism (imprint – desire – action). Automatisms (*saṁskāras*) in the dialectic of life lead to phobias, complexes, and social conflicts. The first lesson of Yoga is to keep the mind (*citta*) from engaging in a variety of sensory-semantic images (*vṛtti*).

Yoga recognizes that between automatism and individual choice there is a gap – an act of volition that can arrest the chain of thought in order to realize the *epoché*. Consciousness merges with the images and automatic performances in the moment of interaction. When mind (*manas*), immersed in its sensual-meaning modification, meets with the new challenge, it treats this as a threat to the current image and enters the affective state (resentment, jealousy, envy, hatred, anger). Affective states are the cause of suffering and delusion, so Yoga teaches us to be in our natural state, to keep the image (*vṛtti*)) subordinate and break the identification with the affective automatisms of consciousness.

Vāsanā are hidden in the subconscious as "seeds of desires," so Yoga considers them to be the most bitter enemies for the *jīva*. *Citta* takes the form of various mental states of factual being. Soul (*jīva*) identifies itself with *citta,* and because of that, the *jīva* enjoys and suffers. Yoga calms *citta*, going through 5 levels of spiritual life, using the 8 stages of Yoga practice, which are based on *tapasya*, asceticism. Thus, a person "anneals" the seeds of desires and attains the state of freedom, happiness, and ultimately liberation (*mokṣa*). When all the imprints of past experience (*saṁskāras*) disappear, the yogi leaves his body and enters the *videha mukti*, liberation after death, in which Īśvara (God) helps him.

Yoga-sūtras (1.3) suggests that one leave everything in the natural state ("*tada drashtuh svarupe avasthanam*" – the seer rests in his own state). One just needs to remove the noise from the mind. Memory is responsible for storing imprints of sensory experience. Patañjalii offers two types of meditation to deal with *saṁskāras*. The first meditation is to reflect on the true ego, full of knowledge, existence, bliss. The second practice is to retain the thoughts (*epoché* – *Yoga-sūtras* 1.17-18). This type of meditation is close to Tibetan Buddhism, where the attainment of *nirvāṇa* is conceived as the gradual extinction of the flow of thoughts and desires:

> Whatever you see around you – all this is a naturally occurring manifestation and a naturally empty void: leave everything as it is, do not create anything with the mind. Thoughts will disappear themselves, if you do not cling to them. Naturally released manifestations become your assistants, and you can practice by adopting external vision as a path. As for what's inside, then, whatever movement may occur in your mind, no matter what thoughts may come – all this has no effect, all is emptiness. Thoughts are released naturally. If any *kleśa* appears, look at it carefully; it will disappear. All around are your assistants.[3]

A special Husserlian technique for transcending *kleśas* is called "phenomenological *epoché*," which takes effect as a basic *phenomenological solution*. Of course, it takes on its radical meaning only at the last (transcendental) level, but it is certainly also true that without the "drop off" of transcendental claims, no reductive work would be possible. Husserl wrote that the eidetic phenomenologist performs *epoché* with respect to any objectifying "position in the emotional experience, both as regards the examination of individual phenomena, and in respect of the whole structure of psychic life,"[4] but based on the ego *cogito*, retaining the "significance of the apperception of the world. As soon as he radically eliminates this apperception, there is a paradigm shift, which affects his life and his psychological studies."[5]

So, the basic step of transcendental reduction involves switching off apperception of the world and "bracketing" our belief in the existence of the world. Everything that undergoes such a modification does not disappear, but in contrast, is revealed as a phenomenon in its own semantic existence. In general we can say that the transcendental reduction is to turn off all transcendental presuppositions and the ideas about the existence of the world. An instance of transcendental reduction is in detecting meaningful differences, "something as something," – not an identity, but a distinction that is supposed in an intentional experience. To distinguish something as something means to select it in the phenomenal order of givens from its availability in the world within which it takes the predication of the transcendent in-itself, etc. Reducing the thesis of the world, i.e. belief in the existence of the world, we thus reduce the world as the ultimate referent (in regard to meaning, presence, etc.). We do not refer as such to the world, but we ask about it in the "how" of its life and "what" of an entity.

The epistemological aspect of transcendental reduction opens the possibility of clarifying the nature of knowledge and its appreciation. This refers

3 Padmasambhava, "Directing the Stuff at the Old Man," // http://spiritual.ru/lib/padma_posoh.html

4 Husserl E. Phenomenology, article in the Encyclopedia Britannica // Logos. 1991, №1.

5 Husserl E, "Amsterdam Reports," pp. 2, 17.

to the fact that phenomenology acts on the level of law, or the "principle of all principles": to consider what is and how it comes from itself, that is, to perceive the essence of self-accomplishment.

On the level of the study of psychic phenomena, it is the study of emotions and their relationships in a single experiment, based only on what can be found in it evidently and in the pure form, that is outside of thought (or any other) projections on the physical as outside of me and independent of me, yet impacting on me. The dimension of the transcendental experience and phenomenological principle are motivated by significantly different interests. There is a difference between being asked about the nature and structure of certain experiences, in order to ask how in the immanent order of consciousness the object is given, that is, asking about the meaning of intentional relations and its origins.

The ontological aspect of reduction opens a new transcendental sphere of experience as a sphere of absolute being of consciousness, transcendental subjectivity. Thus, the ontological difference is determined by the ways of conscious being (the absolute, immanent, pure) and to the world of consciousness (intentional, meaningful existence). Thus, in the course of the phenomenological experience of the world, the first is the positioning or the natural setting, that leads to the second in the course of argumentation (yet first by meaning) – namely, to the ontic unit of the irreducible sphere of ego *cogito* and the ontological or transcendental entities, that alone determine the possibility of positioning of natural datum and the sense of meaningful existence.

Thus, we have two strategies to "fight" with multiplicity. Man, by virtue of his limitations, is inevitably drawn to these procedures. The first is the depletion of the plurality, the conscious confinement of the manifold. This is the work of self-imprisonment of consciousness, for hell is other people (Jean-Paul Sartre).

The elaboration of the emptiness of the world has a long system-genetics: Taoism, Buddhism, postmodernism, and the physical vacuum in physics. Dvaita Vedānta emphasizes the need to personalize one's relationship with God. To do this, the school recommends one accumulate positive *samskaras* (imprints) regarding contact with God – the worship, contemplation, and discussion of acts of God. With spiritual *saṁskāras*, the vector of natural thinking is always directed toward God: "O Lord of Madhu, as the Ganges forever flows to the sea without hindrance, let my attraction be constantly drawn unto You without being diverted to anyone else." (*Bhāgavata Purāṇa*, 1.8.42)

God is seen in Dvaita Vedānta as the universal form of all individual forms (*akhila-rasāmṛta-mūrti*). The *Bhāgavata Purāṇa* (3.9.11) puts this in terms of metaphysical plasticity: "You are so merciful to Your devotees that

You manifest Yourself in the particular eternal form of transcendence in which they always think of You." A similar statement can be found in the *Brahma-saṁhitā* (55): "I adore the primeval Lord Govinda, the meditators of whom, by meditating upon Him under the sway of wrath, amorous passion, natural friendly love, fear, parental affection, delusion, reverence and willing service, attain to bodily forms befitting the nature of their contemplation."

The Advaita followers' activities of self-examination and cessation of chains of thought that pertain to material objects is rather a general rule for the control of the senses (the mind-*manas*, as the chief of the senses). Such a practice of concentrating the mind on the reflective inner-self is inherent to yogis as well. Followers of Dvaita Vedānta, however, control the mind by directing it toward devotional service to God. In the personalistic school, individuality is of great importance because of the specification that we are dealing with a spiritual plurality, while in the school of Advaita plurality of the world is subjected to a total reduction. For the Dvaita school, the material world may be reduced, while the spiritual diversity is subject to constant specification.

A comparison of phenomenology with the Eastern psychotechniques has allowed us to stress the importance of Husserl's break with the tradition of speculative philosophy. Phenomenology implies a pure experience of self-observation. Such a radical intellectual position correlates well with the phenomenology of Yoga and Tibetan Buddhism, which are also built on a break with past tradition. They all reformat within the cultural norm in order to reset and justify the experience of transcendental meditation.

The achievement of phenomenology is not only in its reformist trend in philosophy; it also solves the problem of overcoming the current crisis within the social sciences in Europe by breaking the deadlock of speculative philosophy and positivism. Its universal value is in the development of cross-cultural research methodology by which such systems as Vedānta, Yoga, Buddhism, and phenomenology itself can reside in a single intersubjective field.

Advanced Astronomy in *Bhāgavata Purāṇa*

Tiziano Valentinuzzi
Bhaktivedanta Institute for Higher Studies

Presented at the
Puranic and Siddhantic Cosmology Conference (Mumbai)
November 4–6, 2022

Here I will present topics discussed in a paper by Dr. Roberto Caimmi, an astronomy professor I worked with at the University of Padua. This paper was originally published and presented in 2018, and today's discussion will briefly summarize some new, related ideas which we are currently researching. These include the so-called flat earth model in the *Bhāgavata Purāṇa*; Dr. Richard L. Thompson's investigations of the mythic vs. scientific nature of Bhū-maṇḍala; connections between Bhū-maṇḍala and Vastu Puruṣa Maṇḍala as well as the Titius-Bode Law; and a few conclusions.

First, we will address the notion that the earth is described as flat in the *Bhāgavata Purāṇa*. Many scholars, especially Western ones, think this is the case. This is because of the descriptions of Bhū-maṇḍala in the Fifth Canto, which appear to describe the earth but also say that the disk of Bhū-maṇḍala is very, very thin compared to the volume of the Brahmāṇḍa, or the whole universe. Bhū-maṇḍala is situated within the *antarikṣa*, or the space between the upper and lower hemispheres of the universe. If we look at the universe as a whole, this *antarikṣa* structure is so thin and flat that it is very hard to see. But we need to take into account that in the *Bhāgavatam* and other *Purāṇas*, it is explained that all the planets are moving within this *antarikṣa*. It therefore makes more sense to think that this model corresponds to the solar system, which means Bhū-maṇḍala would be the ecliptic plane.

This conclusion is found in the work of Dr. Richard L. Thompson, or Sadāpūta dāsa, as explained in his book, *Mysteries of the Sacred Universe* (MSU). In that book there are details about his research and helpful diagrams showing the seven *dvīpas* within the larger structure of Bhū-maṇḍala. Dr. Thompson amply demonstrates in this book that Bhū-maṇḍala is not simply

Figure 1. Cover of *Mysteries of the Sacred Universe: The Cosmology of the Bhāgavata Purāṇa* (2000) by Richard L. Thompson.

a flat model of the earth, because the dimensions of Bhū-maṇḍala are compatible with the dimensions of our solar system.

In *Mysteries*, Table 4.1 "List of Astronomical Sizes" lists the radii of various planetary orbits in the solar system and other notable measurements from modern data alongside the radii of various features of Bhū-maṇḍala in miles. The radius of Saturn's orbit is 890 million miles and the radius of Lokaloka Mountain is calculated to be 1 billion miles. The radius of Uranus' orbit is 1.79 billion miles and the outer radius of Bhū-maṇḍala is calculated to be 2 billion miles. Meanwhile, the distance to Alpha Centauri is 25.5 trillion miles. In this way, we can clearly see that the dimensions of Bhū-maṇḍala's features are on the scale of our solar system. In Figure 4.6, Dr. Thompson depicts the plane of Bhū-maṇḍala next to that of our solar system as far out as Neptune's orbit, to scale, with a value of eight miles per *yojana* and an angle of view of 75° from the vertical. The two representations look very similar. These arguments convincingly show that the model of Bhū-maṇḍala is completely consistent with solar system dimensions.

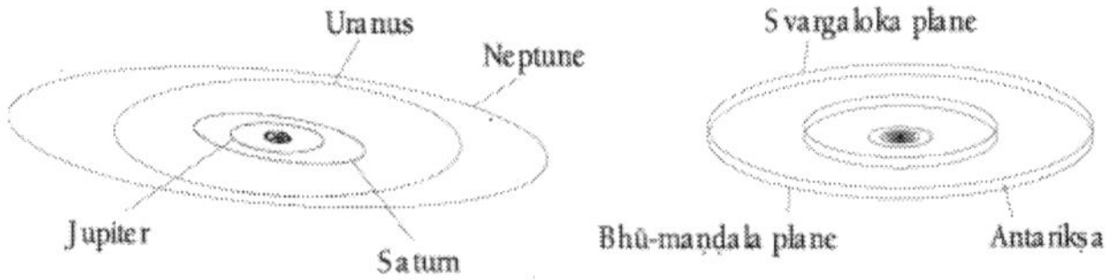

Figure 2. In Figure 4.6 in MSU, Dr. Thompson depicts the plane of Bhū-maṇḍala next to that of our solar system with a value of eight miles per *yojana* and an angle of view of 75° from the vertical. This convincingly shows that the model of Bhū-maṇḍala is consistent with solar system dimensions.

In recent years, I have been examining some qualitative ideas about the nature of complex scientific studies in ancient times. Back then, it was common to create streamlined maps and what I like to call 'infographics' to convey complex concepts. In this context, an infographic is a scheme of information codified in clear, interrelated images and simple relationships, like the doubling feature in Bhū-maṇḍala's concentric islands and oceans. Why was this common? Because such infographics are heuristics – easy to memorize and easy to apply in everyday use. Everyone in India knows about Bhū-maṇḍala because of its presence in the *Itihāsas*, for example the *Mahābhārata*. People might not know the exact details of Bhū-maṇḍala, but everyone has heard of it. On the other hand, only distinguished scholars know the *ślokas* of the *Siddhāntas* – the common people have no familiarity with them at all. My professor's theory was that Bhū-maṇḍala and many other infographics like it came from an ancient science which predates the *Siddhāntas* and possibly even the *Vedas*. He referred to the society that produced this science as an antediluvian civilization.

At any rate, even in modern times we see people using stories with exaggerated, formulaic details to aid in memorization of important concepts. This would have been even more important in ancient times because there were no computers, notebooks, or paper to keep records, and everything was calculated and recalled mentally. This is interesting to note because, in Western countries especially, there are people who say that all these old stories are just invented and not accurate, and thus could be missing the entire point. But it is not about whether these stories are true in every detail, but whether they help people remember important and complex information. Therefore, these stories do not discredit Vedic texts, but actually hint at an advanced science. I have not encountered anything like this in Western scriptures.

So, even if we give these ancient thinkers the smallest benefit of the doubt when it comes to stories, this model implies a very scientifically advanced culture. For me, this is the most important consideration. I also see that for the Vedic and Siddhāntic philosophers it was crucial to link astronomical and human affairs to spirituality. I will discuss this further in my conclusions, but I wanted to stress this because many Vaiṣṇavas today are reading and studying the *Bhāgavatam* with the goal to go back home, back to Godhead. This is always the ultimate aim. Thus, the fact that the aim was similar in ancient times highlights the advanced nature of this culture.

Another example of this kind of story is the Vastu Puruṣa Maṇḍala, which is known for its encoding of architectural principles – one of my personal interests. In the *śāstra*, Vastu Puruṣa was an *asura* who was devouring everything in the universe in uncontrolled anger, including space itself. This is reminiscent of a black hole, which is an "uncontrolled devourer" because it is a zero-point space and everything which enters its event horizon disappears completely. Interestingly, the solution Brahmā adopted was to regulate or control the *asura* with different demigods in all directions, each representing a special energy or influence on space. In other words, the laws of nature must regulate the energy of spacetime, otherwise nothing can be manifested in the universe. This is a very interesting concept that seems to express complex ideas of how the universe came into its current form.

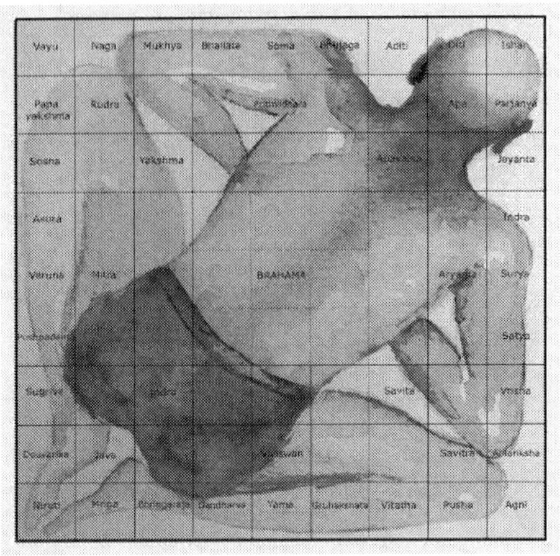

Figure 3. Vastu Puruṣa Maṇḍala: representing the different demigods in all directions (the laws of nature) that regulate the energy of spacetime.

Perhaps we could not call this approach an exact science, but we can see that there are many parallels with concepts in our modern precision science. It is also very significant that these ideas have been passed down to us, because accomplishing this kind of information transfer is not a trivial problem. How does one design packages of complex information that can survive over centuries unadulterated like that? It is an extraordinarily difficult task which is nevertheless being accomplished by the *Vedas* and the *Purāṇas*. For the Vastu Puruṣa Maṇḍala, the idea is that you have all of the design information you need to construct a building contained in an infographic. The importance of Northeast, the delicate condition of Southeast and Northwest, how ability rules in Southwest, keeping the center of the structure open because Brahmā is there, and keeping Northeast light while Southwest should be heavy – these are all the guidelines of *vastu* in one simple scheme.

This leads to one of my mantras: "ancient" does not mean "antiquated." Ideas are not necessarily obsolete or irrelevant simply because they were first expressed centuries or millennia ago. For example, there is a great propensity towards synthesis – combining diverse concepts into a coherent whole – in the *Purāṇas*: synthesis instead of analysis, which denotes an advanced culture, not an antiquated one. Nowadays, we are in the era of "big data," with millions of precise data points being recorded by cutting-edge technologies, but few can synthesize all of that data to come up with useful conclusions. In Sanskrit terms, we have a great deal of *pratyakṣa*, very little *anumāna*, and no *śabda* in the Western world. Where is the real advancement of our society?

I will relate a personal story that illustrates this perspective. I was diagnosed with diabetes about five years ago. In the last few years, I have been given a great deal of speculative advice for this health issue based on people's direct experience or inference from existing scientific information. One doctor told me, "You will be on insulin for the rest of your life. You will never heal. All of your pancreatic cells are gone." Others said, "Eat fruits," "Don't eat fruits," "Start a keto diet," "The keto diet is dangerous for vegetarians," "Become a crudist," "Take these pills," "Cook your vegetables," "Eat your rice raw," etc. Everyone I was consulting had different ideas, and it was obviously impractical to try to follow them all. Then, just one week ago, my friend sent me a paper by Dr. Roy Taylor MBE of Newcastle University, Great Britain, which said that the pancreas regains its lost volume and shape when type 2 diabetes is put into remission through a low-calorie weight management program. Two months of weight loss, and your diabetes is gone. This has been proven in hundreds and hundreds of clinical demonstrations.

When I read this, I laughed and said, "Oh, this is *Bhagavad-gītā*." All of this research took 40 years with a team of 55 PhDs, and this conclusion is already in *Bhagavad-gītā* verse 6.16: "There is no possibility of one's becoming a yogī, O Arjuna, if one eats too much or eats too little, sleeps too much or does not sleep enough." So what does this show? The transcendental knowledge in the *Purāṇas* and *Itihāsas* is also very practical. Meanwhile, we are using big telescopes and big microscopes and investing hundreds of years and enormous amounts of effort to come to the same conclusions.

Coming back to cosmology, Richard Thompson cited four different models for interpreting the Bhū-maṇḍala plane: as a planisphere projection, as the abodes of the *devas*, as a map of central India and Pakistan, and as the geocentric orbits of the planets in the solar system. He demonstrated that for a particular length of *yojana* close to 8 miles, all of the geocentric orbits of the planets are somewhat tangent to the features of the seven *dvīpas*. Those who are interested in further details of these models may find them in *Mysteries of the Sacred Universe*.

What my professor and I are proposing is a fifth model which is based on the Titius-Bode Law. This law, which was articulated in J. E. Bode's astronomical

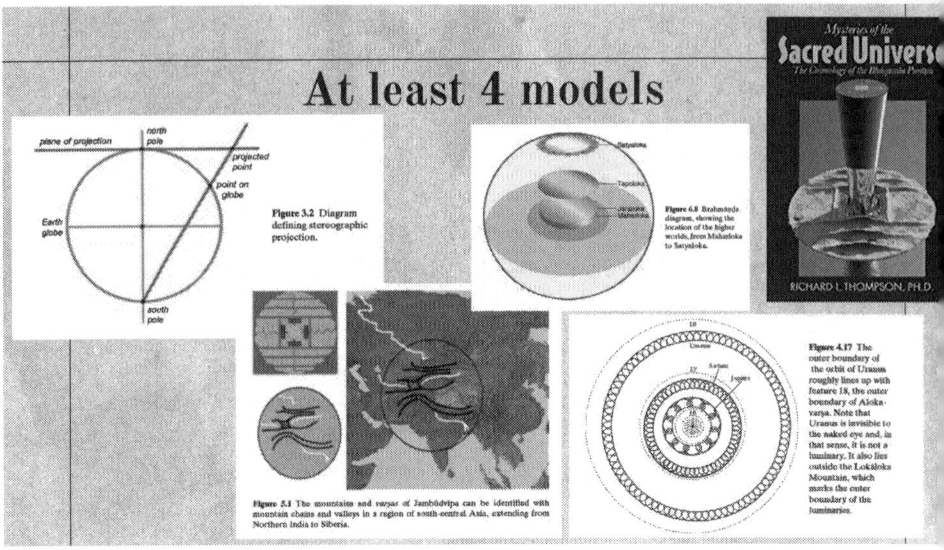

Figure 4. Richard Thompson cited four different models for interpreting the Bhū-maṇḍala plane: as a planisphere projection, as the abodes of the devas, as a map of central India and Pakistan, and as the geocentric orbits of the planets in the solar system.

compendium published in 1772, predicts the distances of planets from the sun by applying a very simple formula with a geometric progression of base two. In other words, extending outward, each planet should be approximately twice as far from the sun as the one before. The formula is $di = 0.4 + 0.3 \times 2i$ given $i = -\infty, 0, 1, 2, 3$ ... where i is the index for each planet (Mercury, Venus, Earth, Mars, etc.). The measurement is given in astronomical units (AU), given 1 AU is about 150,000,000 kilometers or 93 million miles. This law gave good results for the lengths of the semi-major axis for each of the planets known in the 18th century. An important confirmation came in 1781 when William Urschel discovered Uranus, which fell perfectly in line with the law at index $i = 6$. However, at the time there was no planet for index $i = 3$, meaning at the expected distance in between the orbits of Mars and Jupiter. Twenty years later, an Italian astronomer named Giuseppe Piazzi found Ceres. Ceres is classified as a dwarf planet and is the largest asteroid in the asteroid belt. These successful predictions showed that the Titius-Bode Law was doing its job.

An interesting observation that requires more experimental confirmation is that the Titius-Bode Law also seems to work for planetary systems other than our solar system. So, for example, there is a planetary system around pulsar PSR-B1257+12, and all three of its planets have orbital distances which follow the Titius-Bode Law. This was the first positive experimental verification of the extrasolar application.

One caveat is that the Titius-Bode Law stopped being consistently accurate in its predictions for planetary orbits past Uranus. It did not work for Neptune, although it did for Pluto. It seems that trans-Uranian planets and dwarf planets at these greater distances do not necessarily follow the Titius-Bode Law. This is an interesting observation with respect to our *Bhāgavatam's* Sapta-dvīpa. Another point is that at the present day there is no scientific explanation for this law. It is an empirical law, which means that it is a predictive function which matches the data, but no one knows why (similar to quantum mechanics). It could be connected with something called "orbital resonance," which may create empty spaces between orbits with breadths following a geometric progression.

Coming back to Bhū-maṇḍala, we know the measurements of the Sapta-dvīpa from Purāṇic descriptions. The central zone of Jambūdvīpa is 100 kilo-*yojanas* in breadth, meaning 100,000 *yojanas*. Surrounding Jambūdvīpa are oceans and *dvīpas* that double in breadth. This doubling is reminiscent of the Titius-Bode Law because that means it is a geometric progression in base 2. Keeping modern measurements of the planets in mind, let us try to match the Titius-Bode orbital predictions with the features of Bhū-maṇḍala.

Titius-Bode and Bhu-mandala

	i	k	BHU (ky_c)	TB (UA)	Obs (UA)	Feature
Mercury	-inf	0	0	0,4	0,39	
Venus	0	1	100	0,7	0,72	Jambudvipa
Earth	1	2	200	1	1	Plakshadvipa
Mars	2	4	400	1,6	1,52	Shalmalidvipa
Ceres	3	8	800	2,8	2,77	Kushadvipa
Jupiter	4	16	1600	5,2	5,2	Krauncadvipa
Saturn	5	32	3200	10	9,54	Shakadvipa
Uranus	6	64	6400	19,6	19,19	Pushkaradvipa

Figure 5. Titius-Bode Law and Bhū-maṇḍala: The oceans and *dvīpas* that surround Jambūdvīpa doubling in breadth is reminiscent of the Titius-Bode Law, which is based on a geometric progression in base 2.

Let the geometric progression value at each Titius-Bode index i be K. As $i = -\infty, 0, 1, 2, 3, 4, 5, 6$ from Mercury to Uranus, $K = 2i = 0, 1, 2, 4, 8, 16, 32, 64$, respectively. We can see that this geometrical progression is identical to K (given a multiplication factor of 100) for the sizes of the *dvīpas* in kilo-*yojanas*. We can also see that the T-B-predicted distances for each orbit (in AU) are very close to the observed distances of the planets, with less than 5% discrepancy at the most. Mercury does not have a corresponding Bhū-maṇḍala feature, since according to this scheme it should be part of the center, at the zero point (hence $i = -\infty$). Apart from this zero point, the *dvīpas* seem to represent the distances of the planets from the sun.

I want to highlight some additional calculations involving these distances that proved interesting. First, if we assume 1 kilo-*yojana* = 13,820.8 kilometers, that turns out to be one eighth of one degree of the circumference of Earth. This observation is described in one of my professor's papers, as well as in the book, *Mysteries of the Sacred Universe*. Using this value to calculate the astronomical unit in kilo-*yojanas* results in 10,824, which is very close to 108×10^2. We know from the *śāstras* that 108 is a sacred number, and we also know from modern science that 108 is the ratio of the diameter of the moon to the distance between the earth and the moon, as well as the diameter of the sun to the distance between the earth and the sun. This number appears in so many places. Then, if we express the Titius-Bode Law in kilo-*yojanas* instead of AU, we see that the function is divisible by 108: $d_p\,(Ky_c) = 108 \times (40 + \tfrac{3}{10} \times \Delta_{BHU})$. As expected, this formula calculates the width of the *dvīpas* in kilo-*yojanas*. In the appendix of his paper, my professor also provides a nice mathematical representation of 108 as $1^1 \times 2^2 \times 3^3$ and 360 as $2^3 \times 3^2 \times (2+3)$. He is a mathematician, so he can explore the poetry of numbers like Radhakrishnan does in the *Siddhāntas*. He also analyzed the ratio 3:10 as 108:360 – all highly relevant numbers in ancient books of wisdom. This is not so much in the realm of science and mathematical calculations, but it is an indication that the Vedic sages were very creative and artistic with numbers.

Coming back to Bhū-maṇḍala, it is very interesting that the doubling rule only repeats for seven *dvīpas*. After the seventh, Puṣkaradvīpa, there are other notable cosmological features described in the Fifth Canto, but their distances do not strictly follow the geometric progression. As Puṣkaradvīpa corresponds to the semi-major axis of Uranus, this also happens to be the cutoff point for the Titius-Bode Law's predictive power in our solar system. Is this a coincidence? I do not know, but it is very interesting. The next cosmological feature is called Loka, which is not a *dvīpa* but is inhabited, which means this region must have planets. The distance for Loka is described as 15,750 kilo-*yojanas*. As an experimental exercise, I plugged this distance into our modified Titius-Bode formula and the result predicted that there would be some significant planetary body at 47.65 AU. This distance matches the semi-major axis of Makemake, which is the last big dwarf planet in the outer solar system apart from Eris (which has an average distance from the sun of 67 AU, but also has an orbital eccentricity of 0.4). At any rate, practically speaking, we can see that Loka demarcates the outer boundary for all of the planets in our solar system.

At the beginning of this essay, I mentioned that my professor, Dr. Roberto Caimmi, published a paper in 2018 that formed the basis of the more recent research I have been describing here. His work in that paper was to calculate the difference of radii between the oceans in Bhū-maṇḍala, since they are also major features in addition to the *dvīpas*. This was worth researching because this aspect of Bhū-maṇḍala aligns with the concept implied by the Titius-Bode Law that some orbits are not permitted. It is as if a big ocean exists between one planet and the next which precludes the existence of any orbits in between. Why? As I mentioned above, it is most probably because of orbital resonance, but we do not know. The paper focused on calculating the differences in radii, which is a necessary starting point for understanding the whole phenomenon. He began by observing a problem at the first calculation, because there is no ocean in the center of Bhū-maṇḍala – the oceans begin after the first *dvīpa*. He decided that the first difference in radii, ΔR-∞, was 100 kilo-*yojanas*, which is the breadth of Jambūdvīpa. After that, all the oceans are taken into account. The result, of course, is the same strong correspondence we saw between the measurements of the *dvīpas* and the predictions of the Titius-Bode Law, because the oceans are all part of the same Bhū-maṇḍala system.

Here are his results. First, from Mercury out to Venus' orbit is Jambūdvīpa with ΔR = 0.3. Then, the radii differential between the orbits of Venus and earth is the Lavaṇoda Ocean at ΔR = 0.3; between earth and Mars is the Ikṣurasoda Ocean with ΔR = 0.6; between Mars and Ceres is the Suroda Ocean with ΔR = 1.2; Ceres to Jupiter is the Ghṛtoda Ocean with ΔR = 2.4; Jupiter to Saturn is the Kṣīroda Ocean with ΔR = 4.8; and Saturn to Uranus is the Dadhimaṇḍoda Ocean with ΔR = 9.6. These correspondences show at most 5% error compared with actual measurements, which is perfect for scientists. My professor stopped here because the next planet, Neptune, did not follow the Titius-Bode Law. So then I took up the exercise of calculating the difference of radii of Uranus and Pluto to see how the space between them aligns with Bhū-maṇḍala's oceans, since Pluto's orbit matches T-B Law predictions. The radii differential was 20.29, matching up with the Svādūdaka Ocean, and although the error for this ocean relative to actual measurements is larger, it is still only about 6.5%.

The basic conclusion is that Bhū-maṇḍala is not a model for a flat earth, but rather a fairly accurate model of the solar system. By comparing Bhū-maṇḍala's structure with the Titius-Bode Law, we can infer mean planetary distances from the sun. This perhaps suggests that the ancients understood the way planets revolve around the sun. The empirical Titius-Bode Law was not formalized by scientists until the late 18th century, which is far later than the era of the Vedic sages. But this is not about direct comparisons, because

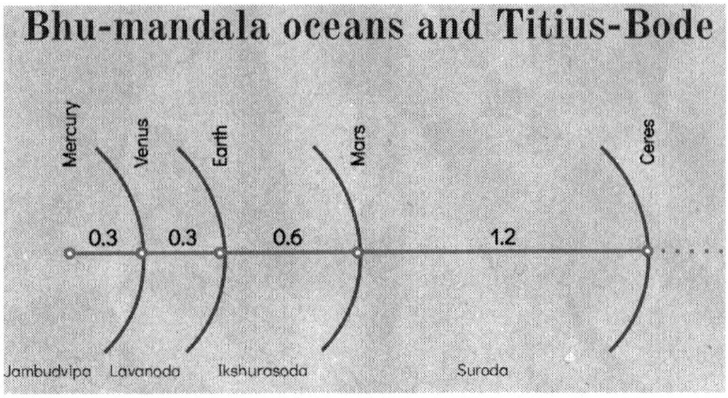

6a

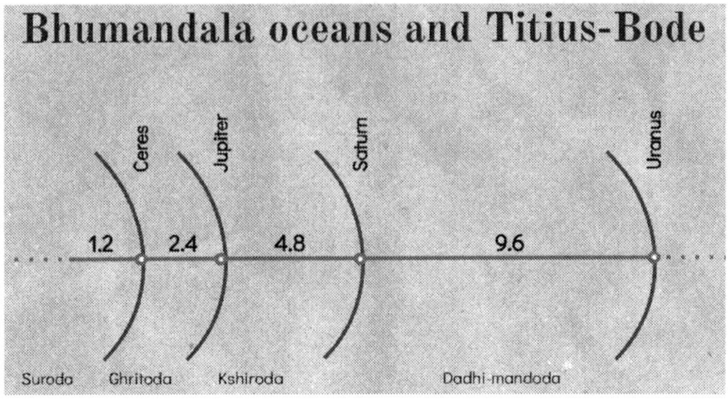

6b

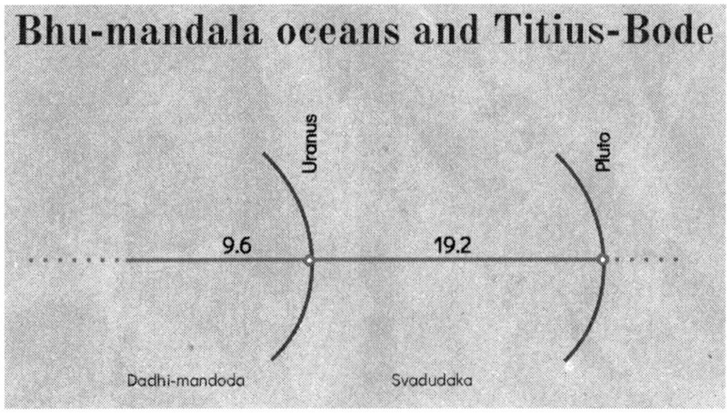

6c

Figures 6a,b,c. Correlating the Titius-Bode Law with the annular oceans of Bhū-maṇḍala.

the two groups had different purposes in dealing with such knowledge. The ancient *ṛṣis* created models of reality which were useful for day-to-day life and not for self-centered activities of theoretical speculation. I am stressing this point again to conclude this essay, because the most important part in these discussions for me is understanding the goal of all of this effort. What is the use of "big data" and the enormous effort invested in obtaining it if we do not try to understand why this universe exists – the purpose of its creation? This is the question underlying all of my research on the *Purāṇas*, which I see as relevant texts for the modern era because "ancient" does not mean "antiquated."

The Vedic thinkers had great concern for connecting all human experience to the divine, because they knew well that this was the only way to find happiness in this life and in the next (as explained in *Bhagavad-gītā*). This is the fundamental recurring point of their works. Perhaps we – both Westerners and those in India – could reintroduce this perspective into modern cosmology and into science in general by considering the spiritual dimension of man as central to our existence. This concept is very important to me, and this surely is the primary message that Vedic culture has been communicating to us from long ago: that we very much need this type of information, ideas, mental models, and ways of thinking, now more than ever.

The Epistemology of Physics and Metaphysics

Howard Resnick
Krishna West, Los Angeles

Presented at the
Consciousness in Science Conference
January 18–20, 2019

To explore the topic of consciousness in science, I begin with a simple, and hopefully uncontroversial, dictionary definition of science: "Science is an intellectual and practical activity encompassing the systematic study of the structure and behavior of the physical and natural world through observation and experiment. Scientists tend to assume that the physical world follows discoverable natural laws and thus behaves within causal relations that are quantifiable, deterministic, and thus predictable." Apart from this, the vlog, "livescience.com," adds that "anything considered supernatural does not fit into the definition of a science." Indeed, the dictionary defines supernatural as "a manifestation or event attributed to some force beyond scientific understanding or the laws of nature."

Thus, to be a science in the "normal" sense of the above definitions, a science of consciousness must focus on deterministic, predictable laws governing consciousness. Here, however, a problem arises: most of us are convinced that we conscious beings possess free will – that human life is not fully deterministic like the life of a machine. This is the exact reason that those branches of science which study unconscious phenomena are called "hard" or "precise," whereas those that study human behavior and consciousness, such as psychology and sociology, are considered "soft," or only approximate.

One thing that philosophy, history and our own consciousness show us, however, is that neither a hard nor a soft science – as science – can ever fully describe reality, including the reality of consciousness itself. This is because science, by its own rules, can only study the physical dimension of an irreducibly bi-dimensional universe that is both physical and metaphysical. Thus, if consciousness ultimately proves to be a metaphysical entity, then neither hard sciences like neurology nor soft sciences like psychology can provide a complete explanation of consciousness. I am to provide clear evidence that a metaphysical realm exists. If it does, what is the metaphysical realm – life's dimension beyond the physical realm?

Guy Kahane, who teaches philosophy at Oxford, asks: "Does God exist? Do we have free will? Are there subjective moral facts? These are familiar metaphysical questions." The Encyclopedia Britannica tells us that "metaphysics was a term used by early students of Aristotle to refer to the contents of Aristotle's treatise on what he himself called 'First Philosophy'... which Aristotle had also referred to as 'theology' (because God was the 'unmoved mover' in his system)."

Stanford Encyclopedia of Philosophy accepts as a metaphysical issue the question of "whether mind is a material mechanism, subject to deterministic laws, or whether it is something else that entails free will." It further notes that the problem of free will (or the problem of the mental and the physical) is among the philosophical problems now considered to be metaphysical in nature, and then makes a key logical point: "Ancient and medieval philosophers might have said that metaphysics was the science that studies being as such, or the first cause... A philosopher who denied the existence of those things would now be considered to be making thereby a metaphysical assertion."

In other words, to affirm or deny a claim or proposition in a particular field of knowledge is to locate oneself – i.e., make a claim – within that particular field. For example, regardless of whether an algebra teacher marks an equation right or wrong, the teacher is making a claim about algebra; regardless of whether one considers a proposed cause of the American Revolution to be correct or incorrect, one is making a claim about history.

Analogously, were a professor at a public university to seriously claim that Zeus is the god of sky and thunder, they would likely be fired for preaching religion. On the other hand, should that same professor claim that Zeus is not the god of sky and thunder, and that there are various neurological and historical reasons that people believed such nonsense, she would be considered a scholar – even though by denying that Zeus is the god of sky and thunder she is equally making a religious claim. Thus, in public universities – and basically all non-religious universities – you are free to preach metaphysical religious positions so long as they are negative, even though both positive and negative claims are within the same domain. This is just one logical point that is routinely overlooked in our incredibly brilliant philosophical age!

Similarly, to claim that consciousness is wholly material is to make not a scientific, but rather a metaphysical claim. One cannot scientifically claim that consciousness is wholly physical because the claim itself is metaphysical in nature. And, by the way, I have nothing against real science; indeed, when scientists are really doing science, I'm actually a great fan. Certainly, rational persons are deeply grateful for all the pain and suffering that science has alleviated. And from a purely intellectual point of view, science has enriched our lives immeasurably and actually saved us from very dangerous, violent forms of religious fanaticism; although sometimes, science itself becomes a dangerous form of religious fanaticism.

In this regard, I will attempt to show that what is called philosophical materialism entails inescapable philosophical and historical problems of great magnitude. This bears on the topic of consciousness since the key issue concerns whether consciousness is simply matter – e.g., a neural epiphenomenon of the brain – or a different kind of thing that is essentially metaphysical.

The highly influential Declaration of Independence declares: "We hold these truths to be self-evident, that all men are created equal, that they are endowed by their creator with certain unalienable rights." Now Jefferson was very intelligent; while once hosting a White House dinner for Nobel Laureates, John Kennedy remarked that this was "the greatest gathering of intelligence at a White House dinner since Thomas Jefferson dined alone." Here we focus on the term "self-evident," which is a key epistemological term, more or less introduced by Aristotle. The idea is that if you make a claim about anything – history, science, God, whatever – people can demand proof; and they can demand that you show proof of the proof. In other words, if one claims that water boils at 100 degrees Celsius and someone doubts this claim, one can place a pot of water on a stove, stick in a thermometer, and allow the doubter to watch as the water boils at 100 degrees. But then that person can question the purity of the water or whether the mercury in the thermometer is real. One must then test the water, then, the water-testing chemicals, and so on and so forth. In this way, one can be pushed into an infinite regress of proofs. To escape this regress, Aristotle teaches that we must take an earnest stand and declare that something is self-evident – that it proves itself.

An example of a self-evident truth that cannot be empirically verified is the claim that there is a real world outside our brain. Now there are certain things that logic forbids. For example, if one claims to have seen a round square, there is no need to empirically test that claim. We know by the meaning of the words that there cannot be a round square; logic forbids it. Descartes, however, raised the point that one can make apparently absurd claims without contradicting logic. For example, the claim that we are simply brains controlled by an evil genius that makes us imagine there is a real world outside of ourselves is certainly very eccentric and difficult to believe, but it does not contradict logic. This shows that without contradicting the rules of logic, one can radically doubt that there is a real word outside one's mind. In modern philosophy, this is called the "brain in a vat" problem: you might be a brain in a vat, being kept alive and hooked up to a super computer.

The fact that we can doubt that there is a real world outside our mind without contradicting logic indicates that in order to do science, one must make a "foundational assumption" (another key epistemological term): one must assert as a self-evident fact that there is a real world outside the mind and that the laws of nature operate more or less uniformly throughout the universe. This assumption serves as an epistemological foundation for empirical science.

However, why do most people, including scientists, assume that there's a real world outside our minds? Because the quality and nature of our experience of the world convinces us that there is no other reasonable explanation. While dreaming, we are absolutely convinced that we are experiencing the "real world," but when we wake up we make a quick comparison and conclude that our waking consciousness is ontologically superior to our dream consciousness. In other words, we consider it to be more real. How do we do that? Why do we do that? There is no way to empirically prove that the world of our waking consciousness is more real than the world of our dream, but we make that assumption because it is self-evident to us that this is indeed the case.

And this is true for every field of knowledge: one must begin with a foundational assumption that something proves itself. Now, of course there are anti-foundationalists; because, in academic philosophy, no one has ever made a claim that at least nine-thousand people didn't try to problematize. But, every anti-foundationalist theory I've seen is itself foundational. Certain forms of philosophical postmodernism, for example, claim that there are no "great truths" – which, of course, is itself a "great truth." Indeed, one of the hallmarks of postmodern philosophy is its great capacity for self-contradiction. Their foundational great truth is that there are no foundational great truths. Ultimately, foundationalism seems to be all-pervading, even among anti-foundationalists, when one closely examines what they are saying.

Here I want to emphasize that just as we make the foundational assumption that there is a real world outside our minds, we also have certain metaphysical views that we hold to be self-evident as well. One such self-evident metaphysical view is that all persons are created equal – an empirically unprovable assumption that forms the basis of democratic society, despite the fact that it is completely contradicted by all empirical science. Democracy, at least in its modern form, is based on the notion of equality, yet science shows that we are not at all equal in terms of beauty, intellectual ability, artistic giftedness, athletic prowess, and so forth. One cannot imagine any empirical test that would prove everyone to be equal. And yet, we have founded our society on this unscientific principle. Why? Because equality is for us a self-evident, foundational metaphysical assumption. And we reject as irrelevant, all empirical scientific evidence to the contrary. Indeed Jefferson himself, in response to Hume's radical skepticisms, declared as self-evident truth that we are all created equal.

Now, one of the main things that David Hume explained, and that basically still stands, is that one cannot derive an "ought" from an "is," meaning that one cannot derive a metaphysical fact from a physical fact. Let's say, for example, that someone commits a non-controversially evil act, such as killing an innocent person with no extenuating circumstances. Analyzing that physical act in terms of forensic evidence, the physiology of the killer and victim, and every other empirical angle will never reveal to us the evil of the act; and yet we hold the act to be evil, one that would traumatize almost anyone that witnessed it for the remainder of their lives. So when we say that such things as murder, genocide, slavery, rape and so forth are evil, we certainly do not mean that blind physical laws of nature and evolution have neurologically wired us to believe in fairy tales. Think of the logical moral implications of such an evolutionary view: genocide is not really wrong; it's just a fairy tale that blind evolution has neurologically wired us to believe. To call an evil act evil is like teaching children about the tooth fairy. In fact, we live in a solely material world in which there are no metaphysical realities. So once we transcend fairy tales, once we stop believing in the tooth fairy, once we stop believing that genocide is wrong, what kind of world will we get?

In his *Critique of Pure Reason*, Kant attempted to defend metaphysics against the onslaught of Hume's skepticism and the powerful, almost hegemonic, rise of science. Unfortunately, his defense basically threw metaphysics under the bus, more or less killing it within the academic world. But even Kant, with his tragic epistemology, believed that our moral intuitions are self-evidently true and cannot be rationally denied. Thus our understanding that genocide is bad, that racism is bad, that it is wrong to murder innocents, is just as epistemically grounded and self-evident as our understanding that we live in a real world, that our senses give us reliable information (something that has been more philosophically contested, and at times refuted, than our moral instincts).

Therefore, in regard to epistemic structure, physics and metaphysics are epistemically parallel. Empiricism has a self-evident foundation upon which we build science. Otherwise, as Aristotle showed, we are pushed into an infinite regress of proofs or we descend into circularity. In parallel fashion, we have a self-evident metaphysical foundation upon which we build our metaphysical systems, such as a national constitution or the laws of the land. Thus, for example, there can be a profound objective sense in which we are equal, and conversely there is a true and objective sense in which racism, sexism and violence toward innocent sentient creatures is wrong. If we want to give that up and claim that the only reality is matter – that murder, rape, thievery and so forth are not objectively wrong – then at least we're being honest materialists. And if we claim that such things are objectively wrong, but still remain materialists, then we're being philosophical hypocrites (apparently not a problem these days).

Previously, I mentioned Darwinism – a theory that has become extremely important in the world. I want to explore just what happens in the real world when you put forward a powerful physical theory while living in a world in which scientists claim that there are no objective metaphysical facts. As I will show, that toxic mixture throws the world out of balance, and leads to disaster. And, by the way, the same thing happens when imbalance occurs on the metaphysical side of the equation, as the history of religion demonstrates.

Looking, first, at the ancient Greco-Roman world, one finds that it was very much syncretistic in its metaphysical views. The Greeks and Romans strongly believed that when encountering other cultures with a different religion, you should do a comparative study and see that the other group is basically saying what you're saying, but with different language, different names for the same metaphysical deities and so forth. The Roman Pliny the Elder, who probably wrote the first encyclopedia, noted long ago that there is actually one reality, but it takes different names. The same view is found in verse 1.164.46 of the *Ṛg Veda*, the earliest Sanskrit text and perhaps the oldest book in the world: "Truth is one, but sages describe it in many ways."

So into this syncretistic, tolerant world (much like our own, with exceptions) enters an extremely fanatical group, which claims that only they worship a living God, whereas everyone else worships dead gods, only they have a true religion, whereas everyone else has a false religion (we've all heard that preaching before). The Romans, as seen in the writings of Tacitus, one of the greatest Roman historians, believed that this type of fanaticism posed an existential threat to culture and civilization. None of this, of course, is meant to justify the cruel persecution of Christians by Romans, which tended to be sporadic and local. In contrast, we have the Christian persecution of pagans after Constantine, which was incessant, categorical and ubiquitous. It continued into the middle ages and even up to the renaissance. The great Renaissance scientist Giordano Bruno said the wrong thing, and was burned at the stake; and Galileo had to make a retraction in order to avoid an attack on both his work and his person.

This violent fanaticism continued into early modernity, and its role in two neighboring countries in the 17th and 18th centuries, France and England, explains much about the course of Western intellectual history. England was often moderate, while France swung to the far right with an absolutist monarchy and church that forbade dissent. For example, in 1685, Luis XIV revoked the Edict of Nantes, which allowed French Protestants to remain in France without being slaughtered.

In England, on the other hand, Elizabeth I (1533-1603) had declared a type of religious tolerance that was quite new in Europe at the time. The Glorious Revolution of 1688 established the rule of parliamentary government over English monarchy. Newton (1643-1727), both the best scientist of his day and deeply religious, played a key role in establishing an independent role for science that was impossible in France. With the Newtonian synthesis – a balance between the physical and the metaphysical – the relative freedom of science and religion both flourished in England.

Back in France, a metaphysical fanaticism – the tyranny of church and divine right monarchy – led to an equal and opposite anti-religious fanaticism in the form of Diderot's hugely influential Encyclopedie, and the murderous French Revolution. In both England and France, we see the clear workings of a dialectical epistemological process, in the well-known form of thesis, antithesis, synthesis. The process is also known as the pendulum effect, or Newton's third law of motion: "for every action, there is an equal and opposite reaction."

Throughout the West's intellectual history, which is deeply affected by political, economic, sociocultural, and military forces, we find that the Western world is merely swinging between two positions. What basically set the pendulum swinging was this intrusion of murderously fanatical metaphysics into what had been in ancient times a syncretistic, tolerant, super-multicultural Europe. During that more tolerant European phase there was a balance between the physical and the metaphysical, and thus most people did not tend to be fanatical in one way or the other. In other words, scientists (or proto-scientists) didn't feel that in order to do their science they needed to bash religion or become philosophically atheist or agnostic; nor did they see religion as the enemy of rational thought, nor insist on strict methodological atheism. For example, the great philosopher Plato rejected the irrational religion of Homer but offered in its place a rational metaphysics. Physics and metaphysics basically worked together to explore different dimensions of reality.

Looking back at the scientific revolution, we find that leaders like Copernicus, Brahe, Galileo and Newton were all religious people. Even Galileo saw the Bible as authoritative in metaphysical matters – i.e. telling how to go to heaven, but not how the heavens go. Many forget that Descartes made an enormous contribution. Today, he is largely remembered as the person who wrongly taught substance dualism – ironically a view that is now making a comeback among a growing number of philosophers.

Historically, when murderous metaphysical extremism arose after Emperor

Constantine, civilization collapsed before a full antithetical reaction could manifest. At that time, the Roman Empire collapsed and people were fighting to survive in very dark, chaotic times. At any moment, one could be invaded, raped, or murdered, and there was little time for philosophical pursuits. Europe plunged into the Dark Ages – literacy almost vanished. In the High Middle Ages, however, Europe began to emerge from the darkness and move toward the Renaissance. Ironically, one of the seeds of the Renaissance was planted by the crusades, which are generally not appreciated. The crusades produced the first major contact that Europe had had with the outside world in a long time.

For example, at the time that Vasco Da Gama landed on the shore of southwest India in 1498, Europeans believed that India was Christian, converted by an apostle of Jesus who miraculously converted the entire population. The Renaissance actually began in the Islamic world, which the crusaders were basically trying to annihilate. Inevitably Crusaders brought back information and culture from this hostile contact with a more advanced culture. Then, the invention of the printing press stimulated growing literacy, because prior to its creation only the rich could own a book, not to speak of a library, since books were expensively hand copied. Then the Protestant Reformation did much to democratize religious belief. This dividing and conquering of religion in the West greatly facilitated the rise of secularism, not least as an antidote to endless brutal European religious wars.

Luther was the most unwilling, inadvertent promoter of modernism one could ever imagine. Rejecting science and philosophy, he had mottos like *sola scriptura*, "only scripture" and *sola fide*, "the only faith." And yet Luther promoted epistemological egalitarianism. Indeed, it was Luther who wanted to get back to the Bible's affirmation of a priesthood of all believers. With the advent of the printing press and Luther's production of a Bible in the German vernacular, all persons could directly read the scripture without having to go through a priest. Prior to that, no ordinary person was able to read the Bible. The first person to make an English translation of the Bible, William Tyndale, was executed in 1536 by Henry the Eighth for revealing a "state secret" (i.e., the Bible).

So with the rise of egalitarian epistemology, people start to read, different opinions arise, and this enormously contributes to an explosion of science – also stimulated by the age of exploration when nations needed faster boats, better weapons, better cartography and so forth. Also, as people read the Bible, many thought, "Wait a second, the priests tell us this, but it is not in the Bible; they have all these mystic sacraments that look like the very same magic for which witches were killed." Thus the search was on for a real science beyond magic, either pagan or Church magic.

This leads to the dramatic rise of science. And when science gets enough power, it goes on the offensive against religion. It's payback time – time for the intellectual, academic persecution of religion. Thus Gibbons reasons that the Roman Empire fell because it became Christian. Marx adds that religion is a drug that makes people stupid and delusional. To this Freud adds that religion is a serious emotional disorder – basically psychopathology. I have a list that goes on and on and on – a war on religion.

And then, of course, science ran into a brick wall in the form of quantum mechanics, with all its problems and weirdness. Quantum physicists routinely describe the most advanced physics as strange, weird, mysterious and so on – the exact language that science used to denigrate magic and religion! And, of course, this admission unleashed a resurgence of magic and religion, as exemplified by 60s books such as *The Dancing Wooly Masters, The Tao of Physics*, and so on.

I compiled a great deal of logical, historical and philosophical evidence for all these points; unfortunately, there's no time to present most of that information here. I will conclude by saying that what's really needed today is an epistemological balance – the restoration of the classical balance between physics and metaphysics. As much as I admire the physical sciences and benefit from their findings in thousands of ways, they are not the absolute gatekeepers of objectivity. They may be the gatekeepers of a certain realm of objectivity and knowledge, but they are not the gatekeepers of all knowledge. Indeed, it is self-evident that we live in a bi-dimensional universe. Even the greatest Western skeptic, Hume, acknowledges that! It is in our Declaration of Independence; it is the foundation of our political system; and it is the foundation of our moral views against sexism and racism. Even prominent atheist philosophers have said, "If there actually are objective moral facts then there must be a supernatural explanation."

My argument is that, given all the facts, it is not possible to be a fully rational materialist; and it is equally impossible to be a rational religious fanatic – religious fanaticism being clearly irrational. We need to get away from fanaticism – whether it pretends to be science or religion. Working together, we should restore the balance between the physical and the metaphysical. The scientific revolution was conducted by scientists that derived essential ideas from their religion. And religious people benefit greatly from reason and science. Let us move forward together along those lines.

The Five Elements of Matter in the *Śrīmad-Bhāgavatam* and the States of Matter in Modern Science

Michael Cremo
Bhaktivedanta Institute for Higher Studies

Presented at the
Nature of Spacetime and the Evolving Universe Seminar
February 19–20, 2022

Of what elements is the world composed? Conventionally, these are the chemical elements, such as hydrogen, helium, oxygen, nitrogen, carbon, and iron. According to Sāṅkhya, as presented in *Śrīmad-Bhāgavatam*, the elements are the *pañca mahābhūtas,* the five elements: ether (*kham*), air (*vāyuḥ*), fire (*analaḥ*), water (*āpaḥ*), earth (*bhūmiḥ*). At first glance these may appear quite different from the elements of modern physics. However, they strikingly resemble the five states of matter recognized in modern science: space, gas, plasma, liquid, solid.

The first Sāṅkhya element *kham*, often translated as ether, appears to correspond to space. In the *Śrīmad-Bhāgavatam* (3.26.34), this description of ether is given:

> The activities and characteristics of the ethereal element can be observed as accommodation for the room for the external and internal existences of all living entities, namely the field of activities of the vital air, the senses and the mind.

Space is generally not considered to be an element in modern cosmology. However, in modern cosmology's most widely accepted theory, the Big Bang theory, the chemical elements cannot form until there is an expansion of space. At the beginning of the Big Bang, matter exists in a superheated, concentrated state, in which atoms cannot form. The rapid inflation of the universe results in cooling, which allows the formation of atoms.

Furthermore, there are scientists and philosophers who say space, or spacetime, is a substance, as in Sāṅkhya. This view is called substantivalism (Maudlin 1993). Some of the supporters of substantivalism say that not only is space a substance, it is the only substance. Material objects correspond to configurations of spacetime. That position is called super substantivalism (Lehmkuhl 2018).

In his commentary on a chapter of the *Śrīmad-Bhāgavatam* (3.11.2) dealing with atoms and time, Śrīla Prabhupāda says, "The atom is the minute subtle form of eternal time." Because all gross material objects are formed from atoms, this statement by Śrīla Prabhupāda could mean that the substance of the material universe is time, perhaps conceived as spacetime. It would be good for a Sanskrit scholar to examine the traditional Sanskrit commentaries on *Śrīmad-Bhāgavatam* (3.11.1–2)

The Sāṅkhya element air can be compared to the gas state of matter. This is straightforward. Just as each of the Sāṅkhya elements is similar to one of the five states of matter, the Sāṅkhya elements and modern science's states of matter manifest in the same sequence. In the *Bhāgavatam*, air develops from ether (space). In modern cosmology, the expansion of space results in formation of atoms and molecules, the first of which are hydrogen and helium gas.

Fire can be compared to the state of matter called plasma. The similarity between fire and plasma is not as obvious as the similarity between air and gas. But the common features of heat and light provide a connection between them. Although some plasmas are cold, most are hot. This is apparent in the following description of plasma on the website of the Plasma Science and Fusion Center at the Massachusetts Institute of Technology (https://www.psfc.mit.edu/vision/what_is_plasma).

> Plasma is superheated matter – so hot that the electrons are ripped away from the atoms forming an ionized gas. It comprises over 99% of the visible universe. In the night sky, plasma glows in the form of stars, nebulas, and even the auroras that sometimes ripple above the north and south poles. That branch of lightning that cracks the sky is plasma, so are the neon signs along our city streets. And so is our sun, the star that makes life on earth possible.

Plasma is often called "the fourth state of matter," along with solid, liquid, and gas. Just as a liquid will boil, changing into a gas when energy is added, heating a gas will form a plasma – a soup of positively charged particles (ions) and negatively charged particles (electrons).

According to this description, plasma is responsible for the visible universe. Visibility requires light and a sense that can perceive it. In this verse from *Śrīmad-Bhāgavatam* (3.26.48), we see the connection between fire and vision: "The sense whose object of perception is form, the distinctive characteristic of fire, is the sense of sight."

In Sāṅkhya, fire is a transformation of air, which corresponds to the gas state of matter. The *Śrīmad-Bhāgavatam* (2.5.27) says

vāyor api vikurvāṇāt
kāla-karma-svabhāvataḥ
udapadyata tejo vai

> "When the air is transformed in course of time and nature's course, fire is generated."

In modern cosmology, hydrogen gas condenses. At a certain density a fusion nuclear reaction is ignited, producing blazing light and heat. So there is a rough equivalence between the two systems.

air	fire
gas	plasma

The next Sāṅkhya element is water, which corresponds to the liquid state of matter. According to *Śrīmad-Bhāgavatam* (2.5.28) it is produced by a transformation of fire:

tejasas tu vikurvāṇād āsīd ambhaḥ

> "Since fire is also transformed, there is a manifestation of water."

In modern cosmology, the heat and pressure accompanying the fusion of hydrogen atoms in the first generation of stars produces helium atoms, and then generation by generation, step by step, the stars produce more elements, going from the lighter to the heavier. One of these elements is oxygen. In stars, an oxygen atom combines with two hydrogen atoms to produce H_2O, water. So in both systems, water comes from fire.

Scientists have detected jets of water being ejected from the poles of stars (van Dishoeck et al. 2021). One of the authors of the study, Lars Kristensen, said, "If we picture these jets as giant hoses and the water droplets as bullets, the amount shooting out equals a hundred million times the water flowing through the Amazon River every second." (https://phys.org/news/2011-06-stellar-super-soaker.html)

The last of the five Sāṅkhya elements is earth, which corresponds to the solid state of matter. *Śrīmad-Bhāgavatam* (3.5.36) states:

jyotiṣāmbho 'nusaṁsṛṣṭaṁ
vikurvad brahma-vīkṣitam
mahīṁ gandha-guṇām ādhāt
kāla-māyāṁśa-yogataḥ

> "Thereafter the water produced from electricity[1] was glanced over by the Supreme Personality of Godhead and mixed with eternal time and the external energy. Thus the water was transformed into the earth, which is qualified primarily by smell."

Earth comes from water. In modern cosmology, the generation of the heavier elements (iron, lead, copper, silver, gold, etc.) that make up most solids comes after the formation of water, which is a combination of two of the lighter elements, the gasses hydrogen and oxygen.

The *pañca mahābhūtas* also have a history in Western cosmology. The Greek philosopher Empedocles (Diels and Kranz 1951–52), said the universe is made of four elements (earth, water, fire, air). Later Greek thinkers added a fifth element, aether, which is sometimes called quintessence, which literally means 'fifth essence." The similarity to the *pañca mahābhūtas* inspires the thought that the Greeks may have gotten their five elements from India. Empedocles was born in a Greek city in Sicily around 494 B.C. and died around 434 B.C. Like Pythagoras he accepted reincarnation and a vegetarian diet, which are signs of possible contact with India. Scholars acknowledge that there were land and sea routes connecting ancient Greece and India, and that people, goods, and ideas traveled these routes (Conger 1952). Among these ideas may have been the Sāṅkhya elements, reincarnation, and vegetarianism. Some scholars object to such transmission. Describing the state of the debate at his time, which arguably still holds today, Conger (1952, p. 127) said: "There is no place for bold assertions, but when the fragments [of PreSocratic Greek philosophy] are viewed in their setting—geographical, commercial,

1 Śrīla Prabhupāda here translates *jyotiṣā* as "electricity". More often he translates it as "luminaries," such as the stars, and this takes us back to the idea that stars produce water.

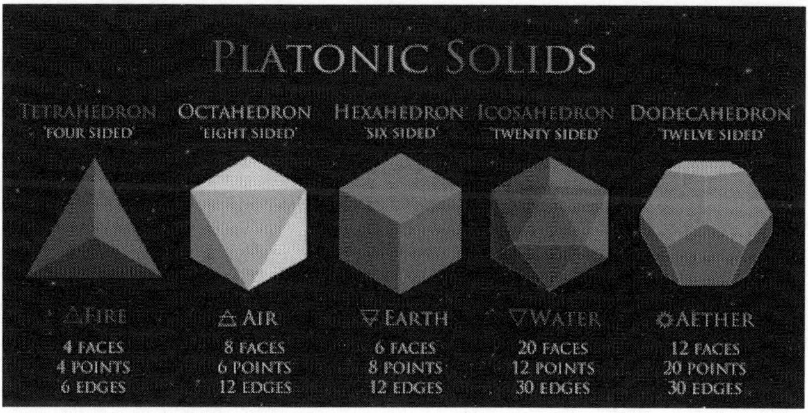

Figure 1: Platonic solids.

and, we may add, logical—the argument for at least indirect Indian influence on Greece acquires some support."

Plato identified the five elements with geometric objects called the Platonic solids. A Platonic solid is a convex regular polyhedron in three-dimensional Euclidean space. Being a regular polyhedron means that the faces are congruent (identical in shape and size), regular polygons (all angles congruent and all edges congruent), and the same number of faces meet at each vertex. There are only five such polyhedra. The first is the tetrahedron, which has four surfaces, taking the form of a pyramid, and represents fire. The second is the octahedron, which has eight surfaces and represents air. The third is the hexahedron, which has six surfaces, forming a cube. It represents the earth element. The fourth is the icosahedron, which has twenty sides and represents the water element. The fifth is the dodecahedron, which has twelve surfaces and represents ether.

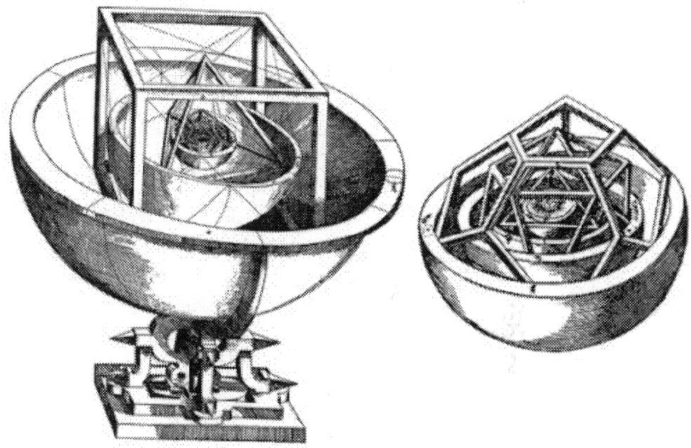

Figure 2: Kepler's nesting Platonic solids.

These Platonic solids were used by the German astronomer Johannes Kepler (1571–1630) in his model of the solar system, published in his book

Mysterium Cosmographicum (1596). He used a nested set of the five Platonic solids to define the orbits of the planets around the sun. He said in the book:

> The Earth's sphere is the measure of all other orbits. Circumscribe a dodecahedron around it. The sphere surrounding it will be that of Mars. Circumscribe a tetrahedron around Mars. The sphere surrounding it will be that of Jupiter. Circumscribe a cube around Jupiter. The surrounding sphere will be that of Saturn. Now, inscribe an icosahedron inside the orbit of Earth. The sphere inscribed in it will be that of Venus. Inscribe an octahedron inside Venus. The sphere inscribed in it will be that of Mercury. There you have the basis for the number of the planets.

In that way, the five elements entered into early modern science through astronomy and cosmology. Interest in the Platonic solids continues until the present. Modern physicists have employed the Platonic solids in their understanding of problems in quantum mechanics, which provides the mathematical foundation of astrophysics (Tavakoli and Gisin 2020):

> Motivated by mathematical beauty and a rich history, we consider the Platonic solids in the context of modern quantum mechanics. Specifically, we construct Bell inequalities whose maximal violations are achieved with measurements pointing to the vertices of the Platonic solids. These Platonic Bell inequalities are constructed only by inspecting the visible symmetries of the Platonic solids.

Because each of the five Platonic solids is identified with one of the five traditional elements, we can see that the *pañca mahābhūtas* have a presence in modern physics. This "fossil" presence of the five Sāṅkhya elements, dating back to ancient Greece, along with the correspondence between the five elements and the states of matter in physics, may provide a foundation for further explorations of connections between Vedic cosmology and modern scientific cosmology.

After noticing the similarities between the *pañca mahābhūtas* and the states of matter in modern science, I wondered who else may have made that connection. A preliminary literature review showed that such comparisons are made primarily by researchers operating from the perspectives of yoga, meditation, New Age spirituality, or paranormal phenomena. Rarely, the comparisons are made by secular scientists or academics, and many of them are from India. The Indira Gandhi National Centre for the Arts, located in New Delhi, sponsored a series of five seminars on the Vedic concept of matter. A five volume series of the proceedings was published (Vatsyayan 1995) and provides a good starting point for further research.

Works Cited

Conger, G. P., "Did India Influence Early Greek Philosophies?" *Philosophy East and West, 2*(2), (July 1952) 102–128. https://doi.org/10.2307/1397302.

Diels, Hermann and Walther Kranz, *Die Fragmente der Vorsokratiker: griechisch und deutsch*, 6th edition, 3 volumes, Berlin: Weidmannsche Verlagsbuchhandlung (1951–52).

Lehmkuhl, D., "The Metaphysics of Super-Substantivalism." *Noûs 52* (2018) 24–46. https://doi.org/10.1111/nous.12163.

Maudlin, T., "Buckets of water and waves of space: Why spacetime is probably a substance," *Philosophy of Science 60*(2) (June 1993), 183–203.

Tavakoli, Armin and Gisin, Nicolas, "The Platonic solids and fundamental tests of quantum mechanics," *Quantum 4* (2020), 293.

van Dishoeck, E. F., Kristensen, L. E., Mottram, J. C., Benz, A. O., Bergin, E. A., Caselli, P., ... & Yıldız, U., "Water in star-forming regions: physics and chemistry from clouds to disks as probed by Herschel spectroscopy," *Astronomy & Astrophysics*, 648, A24 (April 2021).

Vatsyayan, Kapila, *Prakrti: The Integral Vision* (five volumes), New Delhi: Indira Gandhi National Centre for the Arts and D. K. Printworld (1995).

ISKCON's Views on Science and Vaiṣṇavism: Comparing the 1979 *Life Comes from Life* project with the archival transcripts of the recorded conversations

S. E. Kreitzer
Santa Fe College, Gainesville, FL

previously presented at the
ISKCON Studies Institute Conference
Radhadesha Ardennes, Petite-Somme, Belgium
May 22-24, 2015

and the
Bhakti-Yoga & Science Conference
Bhaktivedanta Institute for Higher Studies
Gainesville, FL – November 13, 2016

In 1979, The Bhaktivedanta Book Trust (BBT) published the first English edition of *Life Comes from Life* with the subtitle: *A Startling Challenge to the Modern Scientific Theory of the Origin of Life and the Universe.* The book's publishers described its sixteen chapters as offering "an impromptu but brilliant critique of some of the dominant policies, theories and presuppositions of modern science and scientists by one of the greatest philosophers and scholars of the century, His Divine Grace A. C. Bhaktivedanta Swami Prabhupāda." The Foreword explained that the contents were "based on taped morning-walk conversations that Śrīla Prabhupāda had with some of his disciples during 1973, in the Los Angeles area."[1] Perhaps more than any single project, this work has played a prominent role in helping define Prabhupāda's views about modern science.

As for the original recordings of these discussions, the Bhaktivedanta Archives has made them available along with an organized set of transcriptions since the late 1980s.[2] The audio recordings suggest that the transcribers worked carefully. Yet these transcripts proved significantly different from the text offered in the published versions of *Life Comes from Life*. Whereas the archival transcripts offer a window into a comparatively nuanced discourse promoted as an "impromptu but brilliant critique," the published account can at times appear so different from the recordings that it might be considered a condensed stylized revision if not at points a different conversation. While a close examination confirms a relationship between the book and the recordings, it can at times prove difficult locating even short phrases in the published editions that readily match the archival transcripts. Since the 1979 English edition of *Life Comes from Life* is popularly appreciated as an accurate portrayal of Prabhupāda's discussions with his disciples, a situation of this nature could well present a challenge not only to scholars, but for his mature followers who may wish to develop a comprehensive appreciation of how Śrīla Prabhupāda handled intellectual concerns while promoting the Gauḍīya Vaiṣṇava tradition.

This paper will consider some of the historical context behind the publication of *Life Comes from Life*. While the first edition does not present a definitive historical account, the context for this situation could be sympathetically appreciated when considering various challenges experienced during ISKCON's formative period. This paper will also analyze a few key textual selections to help illustrate the difference between the archival recording transcriptions and the two published editions – first the 1974 German edition, followed by the 1979 English edition based on an English translation of the earlier German manuscript. The argument will be made that the original recordings and archival transcriptions are well worth studying not just for their authenticity, but for their content, as they offer a significantly more substantive portrayal of Bhaktivedanta Swami Prabhupāda's tangible concerns while discussing Gauḍīya Vaiṣṇavism, professional science, and the modern world.

Publication History

The morning walks that formed the basis for *Life Comes from Life* (LCFL) took place during May and December of 1973 in Los Angeles. It is possible that these conversations would have been considered much like previous morning walks if not for the participation of Svarūpa Dāmodara, a young disciple who was completing his Ph.D. studies in Physical Organic Chemistry at the University of California, Irvine.[3] As an example illustrating the significance of these conversations, a published account noted that after these walks took place, Prabhupāda's disciples made it a priority to bring recording equipment on morning walks wherever he traveled.[4] Prabhupāda would encourage his followers to participate,[5] and the devotees who did expressed appreciation that the conversations represented a new initiative.[6]

The following year interest arose to publish these dialogues when BBT editors working in Germany with Hansadutta dasa, the regional Governing Body Commissioner (GBC) based at the ISKCON Schloss Rettershof center, listened to a set of recordings. At the time, Golden Avatar Studios in Los Angeles, a forerunner to the Bhaktivedanta Archives, served as a repository for real-to-real audio recordings that they received from devotees traveling with Prabhupāda.[7] Golden Avatar then advertised subscription options to purchase cassette tapes in ISKCON's monthly journal, *Back to Godhead*. The German devotees likely obtained copies through this service.

Another published account with the title, *Śrīla Prabhupāda and His Disciples in Germany* (1996), had been written by one of the BBT editors who helped produce the initial LCFL German edition. He recalled, "at the beginning of 1974, when we first heard the tapes of Prabhupāda's conversations with the devotee-scientist Svarūpa Dāmodara about the origin of life, we convinced Hansadutta to ask Śrīla Prabhupāda whether we could publish those talks in a book entitled *Life Comes from Life*. Germans are generally scientifically-minded, and we knew it would be great for preaching."[8]

As a GBC leader, Hansadutta proved a dynamic influence increasing book production and distribution. Prabhupāda replied in a letter to Hansadutta dated April 11, 1974: "Yes, by all means you may print my morning walk speaking on 'Life come[s] from Life' into a small book; this argument should be spread, as any intelligent man will be convinced that our discussion is thoroughly scientific and exposes the so called materialistic scientists as rascals. So go on printing and distributing as many books as possible; this is your real work and your personal success."[9]

Life Comes from Life, titled in German as *Leben kommt von Leben*, became one of the German BBT's earliest publications, along with the first volume of Prabhupāda's *Kṛṣṇa Book* (a summary study of the Tenth Canto of the *Bhāgavata Purāṇa*) and the *Teachings of Lord Caitanya* (a summary study of the *Caitanya-caritāmṛta*).[10] The following October, Hansadutta sent published copies for Prabhupāda to review. In reply, Prabhupāda inquired: "And what about 'Leben Kommt Von Leben?' Are they also selling? I have never studied science, but I am challenging them. They may take me as crazy, but I am not crazy, I am right."[11]

Worth noting concerning Prabhupāda's approach – while he would readily admit to not being a scientist himself, he felt qualified to present his challenge based on what he considered essential Gauḍīya Vaiṣṇava principles. In particular, he placed special focus on a "life comes from life" manifesto that consciousness is not an epiphenomenon of matter.[12] He qualified this critique as an exposé of the "so-called materialistic scientists" – as compared to a critique of professional methodologies identified with a scientific discipline. In this context, "materialistic" can be regarded as an adjective describing a philosophical disposition hardly unique to the modern world. Prabhupāda appeared to have considered the popular reputation afforded to the natural sciences as too frequently misused to promote reductionistic philosophical naturalism. When viewing the issue from this perspective, perhaps Prabhupāda was justified in categorizing these proclivities as "so-called science."

As for the production of the 1974 German edition, the text can appear to loosely follow themes in a recorded conversation. It is easy to imagine exuberant young advocates listening to the English tapes while eagerly translating inspirational content they felt could be edited into a dramatic presentation. The flow of the original conversations sometimes became rearranged. Given the pressure to publish quickly with relatively limited editorial resources, this approach may not have appeared a concern at the time. After all, the German edition would at least offer exposure to Prabhupāda's perspectives, whereas prior there was none. As it turned out, the published narrative could to a critical eye appear in parts more reminiscent of a German operetta, than Prabhupāda's directive to "publish a book of lectures" for an English edition.[13]

With the German manuscript complete, Hansadutta wrote to Prabhupāda inquiring about producing an English volume. Prabhupāda replied: "Regarding publishing the *Life from Life* in English it should be grammatically correct because it is [a] written book. Yes, it will be very good if you publish a book of lectures."[14] Of note, Prabhupāda focused his attention on appropriate grammar for a print edition. With that as his criteria, he welcomed a formal publication of his lectures. It is not clear if he read the German edition.

Unfortunately, a raid at the German center the following month by the local police disrupted the project. In the previously mentioned account written by one of the German editors, the author recalled how early one morning, "a well-dressed man in a dark coat knocked on the front door. He was the state prosecutor . . . and he was not alone. Accompanied by eighty policemen, he entered . . . and presented an order of investigation." For "many months," the prosecutor had "gathered apparently incriminating evidence of fraud" involving fundraising done in the name of a food relief program in India, along with related activities.[15]

Though Hansadutta vigorously pioneered book distribution in Germany, aspects of his business acumen could at times lack focus. Prabhupāda commented on this situation in a letter to one of his European leaders: "Regarding dishonest means being used, I have never advised or taught anyone like that. That is not my idea."[16] He confided to another, "This incident in Germany has caused havoc all over the world. It is hampering our reputation everywhere."[17]

While discussing similar concerns with a GBC manager in America, Prabhupāda stated: "Our occupation must be honest. Everyone should adore our members as honest."[18] These disruptive events during ISKCON's early history in Germany could suggest that a conducive environment for mature editorial work may not have been ideal. Suffice it to say, an English edition would need to wait.

In the meantime, the BBT editors in Los Angeles at *Back to Godhead* (BTG) magazine took their own initiative to publish extracts from various morning walks. In early 1974, BTG printed a dialogue with the title, "The Tiny World of Modern Science."[19] A version of this account would also appear in a chapter for another BBT book, *The Science of Self Realization* (1977).[20] The discussion was identified as occurring on April 16, 1973, but the Bhaktivedanta Archives database does not list a morning walk for that date. This could suggest that a recording may have been misplaced sometime prior to the organization of the Bhaktivedanta Archives in the late 1970s. Yet portions of the article can also sound similar to recorded conversations listed in the archives with other dates.

Meanwhile, a second BTG article published in late 1975 shared the same title as the book, "Life Comes from Life."[21] This presentation appears to offer a compilation of excerpts from various sections of an editorial manuscript that would provide the basis for the 1979 English edition. That manuscript, written in English, appears in turn to have been translated from the 1974 German publication – which in its own turn offered a translation of the original English language recordings. Though the text of the BTG article is similar to the 1979 book that shares the same name, the excerpts do not always match exactly. This suggests that the earlier article could have been worked on by different editors, or at least at a different time. One senior BBT editor more recently noted that back in the day, incorporating a degree of license on smaller projects with the intention to emphasize pertinent points could appear a norm.[22]

Though legal investigations into German finances remained a serious challenge during the 1970s, Hansadutta's reputation as an ISKCON pioneer bounced back. During August and September of 1977, Hansadutta received favorable press in Sri Lanka while engaging in a series of debates with a well-known leader of the South Asian rationalist movement, Dr. Abraham Kovoor.[23] Hansadutta appeared to gain an upper hand when he challenged Kovoor to produce life from matter, and if he could not, then he should desist from demanding that religious figures produce definitive proof for God. Hansadutta even offered Kovoor a $60,000 prize if he could deliver. A contemporary BBT account reported that, "Having been challenged [by Hansadutta] to demonstrate that he could create life at a public meeting, Dr. Kovoor declined to attend, saying that he would not feel obliged to demonstrate that he could create life, unless God Himself were to appear in person to demonstrate the same. The [Colombo] *Sunday Times* reported this with the headline 'Flummoxed Kovoor Passes the Buck.'"[24]

Hansadutta regularly sent Prabhupāda updates as the debate unfolded. Even though at the time gravely ill, Prabhupāda expressed enthusiasm for Hansadutta's challenge to the atheistic rationalist. Tamal Krishna Maharaja, who served as Prabhupāda's personal secretary during this period, wrote to Hansadutta on his behalf: "I read your reply [to Kovoor,] and Śrīla Prabhupāda was very pleased . . . Śrīla Prabhupāda stated, 'Hansadutta has become famous now. He is putting very strong arguments. This is preaching.'"[25] In a follow-up exchange, Tamal Krishna Maharaja again reported: "Prabhupāda liked very much your idea to print the talks between His Divine Grace and Svarūpa Dāmodara with an introduction consisting of the correspondence with Dr. Kovoor and yourself. Prabhupāda said, 'It is very good idea. Do it immediately. This will certainly finish the rascal.'"[26]

As in the previous morning walk discussions, Hansadutta presented an argument for life coming from life in sharp contrast to Kovoor's proposition of life as a material epiphenomenon reducible to chemicals. Of relevance to Kovoor's focus as a public intellectual, he had long since retired as a practicing scientist, and now rather indulged in evangelizing atheistic rationalism with a gloss of science drawn from his past experience as a professional biologist. In the end, Hansadutta's success debating Kovoor rekindled interest at the BBT to publish an English edition of *Life Comes from Life*, and now with the added feature of an account of Hansadutta's debate with Kovoor.

The Los Angeles BBT editorial office soon began work on what is considered the classic 1979 English edition. The format of the book reflected the earlier German edition while focusing on the same chapter dates with similar chapter subtitles, albeit now in English. Of note when comparing the texts, the English and German manuscripts appear closer to each other, than to the archival transcriptions of the original recordings. An English editorial draft suggests the same.[27] This situation strongly offers the impression that the English edition draws from a translation of the German edition. This could explain why much of the English publication appears difficult to match precisely with the original English recordings.

An apologetic defense of these circumstances might note that a comprehensive set of recording transcripts would not become available until research commenced on the *Śrīla Prabhupāda-līlāmṛta* biography project that began shortly after Prabhupāda's departure. An account published in the *ISKCON Communications Journal* (2002) about the history of the Bhaktivedanta Archives, offers a description how the BBT decided it would

> fund the transcribing and proofreading of all of Śrīla Prabhupāda's tapes. This would meet two needs: providing material for the *Līlāmṛta* and creating an archival record. It was further proposed to create a microfiche of these transcripts, [in] a project named Operation Vani. A massive typing and proofreading process took place over more than four years, during which devotees typed 40,000 pages of transcripts and finally photographed them onto microfiche.[28]

In 1987, these records were scanned and digitized along with copies of Prabhupāda's published works. These materials subsequently became the digital foundation for what would become the Bhaktivedanta VedaBase program, initially produced in 1991.[29]

Little of all this would have been available during the 1970s when the precursor to the Bhaktivedanta Tape Ministry, Golden Avatar Studios, managed a compact cassette subscription service. Of note, Golden Avatar promotional literature did not mention a back catalogue. In fact, the account in *ISKCON Communications Journal* reported that during this period, reel-to-reel audio tapes could be seen "sitting in an open room and largely unsupervised," with recordings "in large boxes, many unmarked. Many bore no indication of when or where they were recorded." But in 1977, the BBT established the Bhaktivedanta Tape Ministry in order to provide a systematic "catalogue [of] the tapes," which "over the following fifteen years . . . gradually increased" to over 700 recordings.[30] It thus appears that at least until after the late 1970s, it would have proven a challenge to pursue a comprehensive account by locating specific recordings on specific themes. This relative lack of what now in the digital age would be considered a basic research tool, would have made it difficult to appreciate the extent of difference between the initial published accounts and the morning walk recordings.

But even if the editors of the 1979 English edition had been familiar with issues such as these, it may not have been viewed as a priority.[31] Things as they were, the "original" German publication appeared to work quite well as a book, so why not use it as a basis for an English manuscript? To publish would offer something significant. During the following decades, the English language *Life Comes from Life* proved a success. For example, a 2007 edition reported nearly one million copies in print.

A Comparative Textual Analysis

Viewing the 1974 German and 1979 English editions alongside the archival transcriptions of the original recordings helps to illustrate the difference between them.[32] Toward that end, the following section will examine specific selections from a morning walk discussion published as the second chapter in both the German and English editions. The significance for using this sample, is that the Bhaktivedanta Archives offers a recording and transcript that match the date – April 19, 1973 – that is also used for the chapter. The following analysis will focus on text from the beginning and end of the chapter. Sample images of specific selections are included in the Appendix.

The introductory section for the second chapter is represented in four sets of texts: (1) the 1974 German edition based on a translation of the English language recordings [Figure 1]; (2) an editorial draft used for the 1979 English edition, which appears to be a translation of the German edition [Figure 2]; (3) the 1979 English language edition [Figure 3], and; (4) an archival transcription of the original English language recording [Figure 4]. The last image was taken from an encyclopedic set of transcriptions published by the Bhaktivedanta Archives in 1989.[33] The Archives would soon offer search options of digital content as part of its "VedaBase" computer program, and in a service more recently available online.[34]

First under consideration is the beginning of the second chapter published in the 1974 German edition, *Leben Kommt Leben* [Figure 1]. Please note the date at the top of the page (19, April 1973), along with the spelling "Shevot-Hills Park" identifying location. Also note that the first line of text begins with: "Śrīla Prabhupāda: Archäologen ..." Then a little over ten lines down is the phrase: "Śrīla Prabhupāda: Die Drei Erscheinungsweisen der Natur – *sattva, raja* und *tama.*" The Sanskrit terms (*sattva, raja, tama*) refers to the modes of nature – goodness, passion and ignorance – and the Sanskrit is easy to locate in each of the published texts.

The second image [Figure 2] is a sample from an editorial draft used for the 1979 English edition. It came from the files of a Los Angeles office affiliated with BBT production.[35] Please note the same date, and the same incorrect spelling for "Shevot-Hills Park." Also note that the first line at the beginning of the text within the crossed-out section contains the words: "Śrīla Prabhupāda: Archeologists ..." This compares to the first line of the German edition. Then a few lines after the crossed-out section is the beginning of the phrase: "Śrīla Prabhupāda: Three qualities, *sattvas, rajas* and *tamas.*" This matches the Sanskrit terms found approximately ten lines from the start of the German manuscript for the second chapter, as previously discussed.

Now compare these two textual samples, with the first page of the second chapter [Figure 3] of the 1979 English edition of *Life Comes from Life*. Please note once again the same date (April 19, 1973), but now with the correct English spelling for Cheviot Hills Park. Also note the absence of the crossed out section in the editorial draft. The text now begins with the phrase: "Śrīla Prabhupāda: This material world is a composition of three qualities – *sattva, rajas* and *tamas*." These are the same Sanskrit terms found in the German edition ten lines from the beginning, and in the editorial draft after the crossed out section.

These three samples offer textual evidence that suggests the German publication had been used as a basis for the English edition. First, there is the incorrect spelling for Cheviot Hills Park published in the German text (Shevot-Hills Park), carried over to the English editorial draft. Then there is the initial section from the 1974 German edition crossed out in the English editorial manuscript, and subsequently deleted from the 1979 English publication. Finally, the beginning of the second chapter in the English edition contains the Sanskrit terms *sattva*, *raja* and *tama*, as does the editorial manuscript and German publication, albeit later in the text. Of special interest, these three Sanskrit terms are not spoken in the archival morning walk recording that shares the same date.

These examples suggest that after the original 1973 English language recordings were translated into what became the 1974 German publication, the German text was translated back into an English editorial manuscript that provided a basis for the classic 1979 English edition.[36] A track record of this nature might appear less than ideal for producing an accurate account in English.

Along with locating Sanskrit terms, another useful way to identify similarities is by matching verse numbers. A prominent example from the opening section is a reference to "*Bhagavad-gītā* 9.10."[37] This verse citation is found in each text as follows: (1) in the 1974 German edition, in the middle of the second page of the second chapter (p. 18); (2) in the editorial manuscript, in the middle of the second page of "Morning Walk II" (p. 12); (3) in the 1979 English edition, at the top of the second page of "The Second Morning Walk" (p. 8); and (4) on the first page of the archival transcription. Other than the direct reference citation "*Bhagavad- gītā* 9.10", clear similarities do not appear obvious.[38]

While a comprehensive examination of each manuscript can provide examples of a similar nature, the wording is hardly ever exact and frequently placed in a different sequence.[39] This in good part could be explained by the publication history described – the 1979 English edition manuscript may well have been only indirectly connected to the English language recordings, having rather been more directly drawn from a German manuscript subsequently translated back into English.

Next under consideration is the concluding section to the second chapter of the German and English publications, along with the English editorial manuscript that appears to offer a bridge between the two. These three texts will then be compared with the archival transcription that shares the same date: April 19, 1973. Both the published texts and editorial draft conclude with an analogy of a croaking frog inviting death.[40] While these three texts end with this passage, the recording continues for six more minutes. Of note, it is after this passage that Prabhupāda presents a comprehensive summary of his critique. Or to put it another way, Prabhupāda's expression of a central argument had not been included.[41]

In the archival transcription that continues after the published editions stop, Prabhupāda offered an argument that no matter the degree to which material science may take credit for advancing, disconcerting elements of conditional experience remain unchanged. In particular, the quest for lasting youth remained elusive, while the death rate remained steady at one hundred percent. Prabhupāda would also poke fun at those who place their faith in material arrangements while attempting to secure lasting remedies to these primal dilemmas.[42]

Prabhupāda next prodded his disciples: "So you all write very strongly, vibrantly ... Yes. They're misleading. Godlessness." Of note, the audio recording clearly suggests that Prabhupāda spoke the word "vibrantly" at the 47:35 minute mark, and not "vehemently" as misheard by the transcribers.[43] He continued: "As soon as you say 'God created,' immediately they become arrogant. That is our protest." With this, Prabhupāda blatantly asserted his challenge to those who evangelize atheistic metaphysics in the name of modern science. Prabhupāda then proposed: "If they accept God, then we give them all credit. That's all right. Otherwise zero." He emphasized: "We don't deprecate their intention of advancement of knowledge," before concluding, "We simply protest against their defying the authority of God. That is our point."

In these statements, Prabhupāda focused his analysis of material science upon issues pertaining to natural philosophy, rather than professional methodologies identified with a scientific discipline. It is also of interest that within this protest he expressed respect for an aspiration to expand awareness as per his authoritative statement: "We don't deprecate their intention of advancement of knowledge." Rather, Prabhupāda specifically focused on an atheistic interpretation of natural philosophy that indulged in "defying the authority of God." He argued that this was his "simple protest." He further acknowledged that when responding to non-scientific metaphysical beliefs promoted in the name of professional science, that his disciples could cross a line "even [if] it is a little offensive."

Digital copies of the audio recording are readily accessible. The following excerpt, absent from the published editions, begins at 46:25 mark of the Bhaktivedanta Archives recording:

> All these scientists, they discover so many things. Why they did not discover something that he would not die? [That] he would not become old? Where is that discovery? They will say: "Yes, in future."
>
> [But] one man is kicking on your face [now], and you are saying: "Yes, in future, when I shall become strong, I shall kick him [back]." But you are, my dear sir, being kicked now. What you are doing now? "Yes, I'm getting strength by your kicking."
>
> So you all write very strongly, vibrantly. Even [if] it is a little offensive, still these rascals should be taught [a] good lesson. Yes. They're misleading. Godlessness. As soon as you say, "God created," immediately they become arrogant. That is our protest.
>
> If they accept God, then we give them all credit. That's all right. Otherwise zero. We don't deprecate their intention of advancement in knowledge. But we simply protest against their defying the authority of God. That is our point.[44]

Swarupa Dāmodara would then propose a solution: "If they hear *Śrīmad-Bhāgavatam*, then their hearts will be changed." To that Prabhupāda replied, "Certainly," and spoke on the significance of book distribution.

In this notable section, Prabhupāda twice emphasized the word "protest" before concluding with the phrase, "That is our point." This set of statements succinctly articulates Prabhupāda's protest against a philosophical attitude identifiable with atheistic material reductionism – a positivistic worldview frequently promoted on the strength of reputation of the professional natural sciences, and with a particular prominence during the mid-twentieth century. It is important to point out that Prabhupāda's protest focused on metaphysical concerns about which scientific methodology lacked disciplinary authority. These statements suggest that Prabhupāda's challenge to modern science is specifically aimed at evangelical atheists who propagate metaphysical beliefs while hiding behind the public reputation of the professional disciplines. Or to put it another way, Prabhupāda aimed to confront a form of "so-called science."

And yet, this concluding section is omitted from the first editions published for broad dissemination. If considering the published editions with scholarly concern, it is likely better to examine the original recordings along with their archival transcriptions to assist in developing a comprehensive appreciation of Prabhupāda's views. For that purpose, the first editions of *Life Comes from Life* fall short as an authoritative account of Prabhupāda's approach toward the intellectual engagement of Gauḍīya Vaiṣṇavism, professional science, and the modern world.

Works Cited

Das, Kurma-rupa (1999): *The Great Transcendental Adventure: Pastimes of His Divine Grace A. C. Bhaktivedanta Swami in Australia and New Zealand.* Botany, NSW, Australia: Chakra Press.

Das, Ranjit (2002): "Preserving Prabhupāda's Legacy." *ISKCON Communications Journal*, Vol. 10.

Dāsa, Vedavyāsa (1996): *Śrīla Prabhupāda and His Disciples in Germany.* [Benissa]: Bhaktivedanta Book Trust España, 1996.

Goswami, Satsvarūpa dāsa (1991): *Prabhupāda Meditations*, Volume II. Port Royal, PA: GN Press.

Prabhupāda, A. C. Bhaktivedanta Swami (1974): "The Tiny World of Modern Science." *Back to Godhead* Vol. 1, No. 61, pp. 3–9.

Prabhupāda, A. C. Bhaktivedanta Swami (1974): *Leben kommt von Leben: Die Wissenschaftliche Grundlage des Krsna-Bewusstseins.* Schloss Rettershof/i. Ts.: Bhaktivedanta Book Trust.

Prabhupāda, A. C. Bhaktivedanta Swami (1975): "Life Comes From Life: Can Science Create Life in the Laboratory?" *Back to Godhead* Vol. 10, No. 12, pp. 5–9.

Prabhupāda, A. C. Bhaktivedanta Swami (1979): *Life Comes from Life: Morning Walks with A. C. Bhaktivedanta Swami and Swarupa Dāmodara Singh.* Los Angeles: Bhaktivedanta Book Trust.

Prabhupāda, A. C. Bhaktivedanta Swami (1989): *Conversations with Śrīla Prabhupāda: Unedited and Unabridged*, Volume Three. Los Angeles: Bhaktivedanta Book Trust, pp. 326–341.

Prabhupāda, A. C. Bhaktivedānta Swami (2015): *The Complete Teachings*, Folio Infobase Program. Sandy Ridge, NC: Bhaktivedanta Archives.

Singh, Dr. Thoudam Dāmodara (1979): "Śrīla Prabhupāda's Vision for the Bhaktivedanta Institute." In *Perspectives on Bhaktivedanta Institute.* Philadelphia: Bhaktivedanta Institute.

Primary Document Digital Resources

Two-column comparison of 1979 English publication and the 1973 original recording

Published version of the "Second Morning Walk," alongside April 19, 1973 recording transcript.

The author has produced a two-column document providing a side-by-side comparison of the 1979 English publication next to the 1973 recording transcript, to help illustrate the sharp contrast between the original conversation recording and the published version of it. See fn. 32 for more information.

The two-column document is viewable at the following address:

https://bit.ly/LCFL_2text_comparison

Original recording and archival transcript (*1973*)

For a link to the morning walk conversation recording and transcript for the date April 19, 1973, used as an inspiration for the published versions of "The Second Morning Walk," see: Bhaktivedanta VedaBase, "Morning Walk at Cheviot Hills Park, April 19, 1973, Los Angeles" (accessed 25 July 2023):

https://vedabase.io/en/library/transcripts/730419mwla/

This website offers a link at the beginning of the transcription for listening to the fifty-three minute recording. The transcription of the recordings appears accurate, short of an occasional typographical error. For example, at one point the transcribers mistook the work "vehemently," for "vibrantly." The latter word can be clearly heard at the 47:35 minute mark of the recording.

Compete digital copy of Leben kommt von Leben (*1974*)

A digital copy of the text for the 1974 German edition *Leben kommt von Leben* is available at the following site (accessed 25 July 2023):

prabhupada.de/Prabhupada%20-%20Leben%20kommt%20von%20Leben.pdf

This is the initial BBT publication based on the 1973 morning walk conversations.

Complete editorial manuscript used for the 1979 English edition

A copy of a complete editorial document utilized for the 1979 English edition can be viewed at the following site (accessed 25 July 2023):

http://surrealist.org/jpgspdfs/lcfl.pdf

See pages 11–25 of the pdf document for "Morning Walk II," which corresponds to the published "The Second Morning Walk" that took place on April 19, 1973. See fn. 35 for additional historical background information for this document.

Complete digital copy of Life Comes From Life (1979)

The following site offers a digital copy of the text to the 1979 English edition of *Life Comes from Life* (accessed 25 July 2023):

https://yoga.krishna.com/en/pdf/e-books/Life_Comes_From_Life.pdf

While the Foreword suggests that the work was drawn from morning walk recordings identified with the dates for each chapter, evidence suggests the English edition drew heavily from a translation of the earlier German first edition.

Digital copy of Figures 1-4 offered in Appendix

The author produced a pdf file containing the four documents [Figures 1-4] offered in the Appendix. See fn. 32 for more information. This document is also available at the following address:

https://bit.ly/LCFL_4documents_comparison

Appendix Figure 1

1974 German publication – 1st page 2nd chapter

Morgenspaziergang
am 19. April 1973
im Shevot-Hills Park in Los Angeles

Śrīla Prabhupāda wird begleitet von Svarūpa-dāmodara dāsa brahmacārī, Karandhara dāsa adhikārī, Brahmānanda Swami und anderen Schülern.

Der Darwinismus ist hinfällig

Śrīla Prabhupāda: Archäologen werden auch als Wissenschaftler angesehen?

Svarūpa-dāmodara: Ja.

Śrīla Prabhupāda: Sie ziehen ihre Schlußfolgerungen nach Knochenfunden?

Svarūpa-dāmodara: Ihr Forschungsgebiet ist die Vergangenheit. Sie versuchen anhand von Überresten der vergangenen Jahrtausende herauszufinden, was früher gewesen ist.

Śrīla Prabhupāda: Sie beschäftigen sich nur mit ein paar tausend Jahren – wir dagegen haben Beschreibungen von Geschehnissen, die Millionen und Abermillionen von Jahren zurückliegen.

Svarūpa-dāmodara: Śrīla Prabhupāda, gibt es auch unter den niederen Formen des Lebens dämonische Arten?

Śrīla Prabhupāda: Die drei Erscheinungsweisen der Natur – *sattva, raja* und *tama* – Güte oder Reinheit, Leidenschaft und Unwissenheit – wirken überall, denn die materielle Welt ist eine Zusammensetzung dieser drei Eigenschaften. Sie sind demnach bei allen Lebensformen zu finden – nur in unterschiedlichen Mischungsverhältnissen –, selbst in den Bäumen. Deshalb trägt der eine Baum wunderbare Früchte, wohingegen ein anderer nur Brennholz liefert, er ist ansonsten nutzlos. Solche Unterschiede beruhen eben auf den unterschiedlichen Erscheinungsweisen, in denen sich ein Lebewesen befindet. Unter den Tieren gilt das gleiche Prinzip. Eine Kuh beispielsweise befindet sich in der Erscheinungsweise der Reinheit, denn sie besitzt vornehmlich gute Eigenschaften; ein Löwe ist voller Leidenschaft, und ein Affe ist unwissend. Darwins Vater war wohl tatsächlich der Affe, denn er stellte nur törichte Theorien auf. Wie kam er überhaupt auf seine Evolutions- und Selektionstheorie?

Appendix Figure 2

Late 1970s editorial manuscript based on German edition – 1st page 2nd chapter

MORNING WALK II

recorded on the 19th of April, 1973,

in Shevot-Hills Park, Los Angeles.

Śrīla Prabhupāda is accompanied by Dr. Singh, Karandhara dāsa Adhikārī, Brahmānanda Swami and another students.

DARWINISM EXTINCT

ŚRĪLA PRABHUPĀDA: Archeolists reach conclusions by seeing tools.

DR. SINGH: Their business is to find remnants of the past one thousand or two thousand years.

ŚRĪLA PRABHUPĀDA: They are concerned with only a few thousand years, but the *Vedas* give descriptions of ages covering millions and billions of years.

DR. SINGH: Śrīla Prabhupāda, not only amongst the humans, but also amongst the lower forms of species, one finds the demoniac qualities.

ŚRĪLA PRABHUPĀDA: Three qualities, sattvas, rajas and tamas (goodness, passion and ignorance) are working everywhere, because this material world is a composition of these three qualities. These three qualities must be present in various proportions in all species, even in the trees. Therefore you see some trees producing nice fruits, while other trees are simply meant for fuel. This is due to the

Appendix Figure 3

1979 English publication – 1st page 2nd chapter

The Second Morning Walk

Recorded on April 19, 1973,
in Cheviot Hills Park, Los Angeles.

Śrīla Prabhupāda is accompanied by Dr. Singh, Karandhara dāsa adhikārī, Brahmānanda Svāmī and other students.

Darwinism Extinct

ŚRĪLA PRABHUPĀDA. This material world is a composition of three qualities—*sattva, rajas* and *tamas* (goodness, passion and ignorance)—which are working everywhere. These three qualities are present in various proportions in all species of life. For example, some trees produce nice fruit, while others are simply meant for fuel. This is due to the association of particular qualities of nature. Among animals also, these three qualities are present. The cow is in the quality of goodness, the lion in passion, and the monkey in ignorance. According to Darwin, Darwin's father is a monkey. [Laughter.] He has theorized foolishly.

DR. SINGH. Darwin has said that some species become extinct in the struggle for survival. Those which are capable of surviving will survive, but those which are not will become extinct. So he says survival and extinction go side by side.

ŚRĪLA PRABHUPĀDA. Nothing is extinct. The monkey is not extinct. Darwin's immediate forefather, the monkey, is still existing.

KARANDHARA. Darwin said there must be a natural selection. But selection means choice. So who is choosing?

ŚRĪLA PRABHUPĀDA. That must be a person. Who is allowing someone to survive and someone to be killed? There must be some authority with discretion to give such an order. That is our first proposition. Who that authority is, is explained in

Appendix Figure 4

1989 Bhaktivedanta Archives publication of recording transcript with same date as German and English edition 2nd chapter

Morning Walk

Los Angeles, April 19, 1973

Prabhupāda: There is such proposition that He should exist, He should not exist. Then there must be some authority to give such order. That is answered in *Bhagavad-gītā*. Kṛṣṇa says: *mayādhyakṣeṇa,* "Under My superintendence, nature is working." *Harer nāma harer nāma harer nāma...*
Svarūpa Dāmodara: He also says that the species are not created independently. They say it is descended from, one after another.
Prabhupāda: So, if there is no question of independence, then how, abruptly, he can begin from a certain species? You must explain wherefrom this species came into existence. *Harer nāma harer nāma...*
Svarūpa Dāmodara: There are so many books on Darwin's theory. Goes... In the library, if one goes there are hundreds of volumes of books on Darwin's theory.
Prabhupāda: They have accepted or protested?
Svarūpa Dāmodara: Mostly they accept and there are some who are also very critical, what they're, what he's saying is really (indistinct) But those are very few. Mostly, most books are supporting.
Karandhara: Recently, there was an issue that some people wanted that the theory that God created the earth and the species to be taught in schools along with Darwin's theory.
Prabhupāda: Yes.
Karandhara: But the, it was defeated because the scientists said: "If we make such a statement in our schools, everyone will take us as fools."
Prabhupāda: What is that? I could not follow. Eh?
Svarūpa Dāmodara: The evolution...

Endnotes

1 Prabhupāda (1979), p. ix. The following site offers the text from the English edition of *Life Comes from Life*: https://yoga.krishna.com/en/pdf/e-books/Life_Comes_From_Life.pdf, access date 25 July 2023.

2 While cassette tapes of recordings had been available earlier, the Bhaktivedanta Archives began cataloguing the audio recordings and producing transcripts soon after it was established in the late 1970s. The transcripts were published in an encyclopedia set late 1980s, and then a few years later offered in a searchable VedaBase computer program. This program is now freely available in an online version. For more information on the VedaBase program, see: www.Prabhupāda.com/Vedabase/VedaBase.html.

3 Singh (1979), p. 11. In 1979 Swarupa Dāmodara published an account of the early history of the Bhaktivedanta Institute, in which he recollects Prabhupāda speaking about the significance of these morning walks for the formation of the Bhaktivedanta Institute. Swarupa Dāmodara wrote that Prabhupāda told him: "It is for this that I trained you on the morning walks at Venice Beach in Los Angeles. That is why I took you every morning at that time." (p. 11)

4 Kurma-rupa das (1999). In Chapter 9, in the section subtitled, "Saturday, 29 June 1974 – Ratha-yatra," the author writes: "Up until recently, only isolated, informal attempts had been made to permanently record Śrīla Prabhupāda's gem-like morning walk conversations on audio tape. It had only been since Prabhupāda's famous 'life-comes-from-life' morning walk conversations in Los Angeles in April 1973, however, that regular, official recordings had begun."

5 Goswami (1991), pp. 63–64.

6 Howard Resnick (Hridayananda Das Goswami) email to author, 18 May & 19 May 2015. Resnick participated in a number of morning walks during the December 1973 sessions, as acknowledged in both the recording transcripts and the published edition of *Life Comes from Life*.

7 In an oral history interview with Ramesvara das held in 1979 as part of the BBT research for the *Śrīla Prabhupāda-lilamrta* biography project, Ramesvara, who served as a prominent BBT manager during the 1970s and early 1980s, noted that the initial donation for the purchase of recording equipment given to Golden Avatar Studios came from Alfred Ford (Ambarisa dasa), the great-grandson of Henry Ford. Alfred Ford would become a formally initiated disciple in 1974.

Ramesvara recollects: "And Alfred Ford gave a donation and Golden Avatar was started in 1972. We bought the Uher tape recorders, we bought whatever equipment there was, you could interview Krsna Kanti ... This was a monumental thing." See: Ramesvara das, "Ramesvara das interview 1979," pp. 204–205, www.govindadasi.com/uploads/1/8/7/4/18743708/ramesvara1979-preview.pdf, accessed 25 July 2023.

8 Vedavyasa dasa (1996), p. 338.

The title page of *Leben kommt von Leben* (1974) named two editors who worked on the German translation: "Vedavyasa dasa brahmancari (Christian Jansen)" and "Harernamananda dasa adhikari (Hartwig Dalldorf)." Accounts from the book and elsewhere suggest both editors were in their early-mid twenties at the time, with only a few years of experience as practitioners while serving as BBT editors.

Christian Jansen remains active within ISKCON as a prominent BBT editor and manager. He took to the renounced order in 2014 and received the *sannyāsa* name Bhakti Gauravani Goswami. In an email exchange with the author (15 April 2015), Jansen wrote that he was not aware of the historical details of the 1979 English language LCFL project, other than that it came out after the German edition. For more information on Bhakti Gaurvani Goswami, please see: https://iskconsannyasa.com/sannyasa/Bhakti_Gauravani_Goswami, access date 12 September 2024.

Meanwhile, Hartwig Dalldorf maintained a close working relationship with Hansadutta (Hans Kary) through at least the early 1980s. Under Hansadutta's direction, Dalldorft served as a senior manager for the "Mt. Kailash" Krishna farm outside San Francisco. That community would also run into legal difficulties. See: John Hubner and Lindsey Gruson, *Monkey on a Stick* (San Diego: Harcourt Brace Jovanovich, 1988), 237.

9 Bhaktivedanta Swami to Hansadutta, 21 April 1974, https://vanisource.org/wiki/740421_-_Letter_to_Hansadutta_written_from_Hyderabad, access date 25 July 2023.

10 A digital copy of Leben kommt von Leben is available at the following site: http://www.Prabhupāda.de/Prabhupāda%20-%20Leben%20kommt%20von%20Leben.pdf, accessed 25 July 2023.

11 Bhaktivedanta Swami to Hansaduta das, 12 November 1974, https://vanisource.org/wiki/741112_-_Letter_to_Hansadutta_written_from_Bombay, access date 25 July 2023.

12 He rather argued the opposite – that matter is an epiphenomenon of consciousness, with the Supreme Personality of Godhead as the ultimate source.

In the archival transcript for the Los Angeles morning walk recording dated May 3, 1974, Prabhupāda offered the following explanation for his use of the phrase "life comes from life." See: Bhaktivedanta VedaBase, "Morning Walk, May 3, 1973, Los Angeles," https://vedabase.io/en/library/transcripts/730503mwla/, access date 22 December, 2023:

> This is a fact ... that life comes from life. ... They are searching after the original source. That is life, Kṛṣṇa, *sarva-kāraṇa-kāraṇam* (Bs. 5.1), cause of all causes. So this is not theory, this is fact. Now we have to prove it. ... This will be our work. Our worshiping of Kṛṣṇa, that is our internal affair. The external affair – we need to establish this theory.

> Prabhupāda would add the following context as he spoke:
>
> In another sense, life does not come; life is existing. It is not exactly the word that life generates. No. Life is existing. The matter is generating. Matter is generating, and it stays for some time, again it is vanished. Just like this body. This body is born at a certain date, and it will be finished at a certain date. This is matter. ...
>
> But life has no such change. When the life is within the matter, it appears that it is taking birth or death ... [but] life has no change. The change is outward, material things. In all respects we have to prove that life does not come from matter. Matter generates from life, stays for some time, and it is finished. Again begins another chapter.

13 Bhaktivedanta Swami to Hansaduta das, 14 November 1974, https://vanisource.org/wiki/741114_-_Letter_to_Hansadutta_written_from_Bombay, access date 25 July 2023.

14 Bhaktivedanta Swami to Hansaduta das, 14 November 1974, https://vanisource.org/wiki/741114_-_Letter_to_Hansadutta_written_from_Bombay, access date 25 July 2023.

15 Vedavyasa Dasa, p. 342.

16 Bhaktivedanta Swami to Ajita das, 6 January 1975, https://vanisource.org/wiki/750106_-_Letter_to_Ajita_written_from_Bombay, access date 25 July 2023.

17 Bhaktivedanta Swami to Brahmananda Swami, 7 January 1975, https://vanisource.org/wiki/750107_-_Letter_to_Brahmananda_written_from_Bombay, access date 25 July 2023.

18 Bhaktivedanta Swami to Rupanuga das, 9 January 1975, https://vanisource.org/wiki/750109_-_Letter_to_Rupanuga_written_from_Bombay, access date 25 July 2023.

19 Prabhupāda (1974). See also page three of the pdf copy of BTG #61 (1974) offered at: http://vanimedia.org/w/images/4/45/090_1974_01-61.pdf, access date 25 July 2023.

20 The essay was presented as a subsection of Chapter 6 in *The Science of Self-Realization*. See: https://vanisource.org/wiki/SSR6g_The_Tiny_World_of_Modern_Science, access date 25 July 2023.

Of note, whereas the BTG article identifies as its source a morning walk dated April 16, 1973, the online presentation of the BTG article identifies the conversation as having taken place on May 15, 1973. Meanwhile the introduction spoken on the recording associated with the "May 15" link, states May 5, 1973. This situation helps to illustrate various difficulties that can be experienced while attempting to match published essays with archival recordings.

Of further note: *The Science of Self Realization* (SSR) version offers additional text at the beginning of its presentation as compared to the presentation in BTG. The text from the BTG article begins nine pages into the text published in SSR (on page 220 of the 1977 edition.) Otherwise, the two presentations match each other, though neither appear to readily match a specific archival recording (as per research at the time of this essay's publication).

21 Prabhupāda (1975). See also page five of the pdf copy of BTG 10.12 (1975): http://vanimedia.org/w/images/6/6d/108_1975_10-12.pdf, access date 25 July 2023.

22 The BBT editor, Dravida Das, discussed historical context regarding editorial standards for a number of early ISKCON projects involving small book compendiums of Prabhupāda's teachings, in a recorded conversation held with the author on May 13, 2015, in Alachua, Florida.

23 For a sympathetic biographical account of the life of Dr. Abraham Kovoor, see: https://arumugam.tripod.com/kovoor.htm, access date 25 July 2023.

24 Prabhupāda (1979), p. 135. The March 1978 edition of *Back to Godhead* magazine (13.3) published an account of the exchange between the two, available at: www.backtogodhead.in/prove-it-by-hamsaduta-swami-and-mahakanta-dasa/, access date 25 July 2023.

The original article can be viewed beginning on page twenty-five of a pdf of BTG 13.3 at the following address:

https://issuu.com/krishnagames/docs/132-back-to-godhead-volume-13-number-03-1978 , access date 25 July 2023.

Hansadutta also offered an account (posted on July 4, 2011) on his personal website at the following address: http://hansadutta.com/krsnaworld/2011/07/04/the-challenge/, access date 25 July 2023.

25 Tamal Krishna Goswami to Hansadutta Swami, 14 August 1977. See: Prabhupāda (2015).

26 Tamal Krishna Goswami to Hansadutta Swami, 10 September 1977. See: Prabhupāda (2015). Kovoor would pass away near exactly a year later, on September 18, 1978.

27 See fn. 35 for additional historical information on the editorial draft.

28 Ranjit Dasa, "Preserving Prabhupāda's Legacy." *ISKCON Communications Journal* 10 (2002). See also: https://www.iskconcommunications.org/iskcon-journal/vol-10/preserving-Prabhupāda's-legacy, access date 25 July 2023.

29 The Bhaktivedanta Archives would publish compilations of documents in encyclopedic sets beginning in the late 1980s. The morning walk conversations were part of a twenty-volume compendium with

the title, *Conversations with Śrīla Prabhupāda—Unedited and Unabridged: Complete Archival Collectors Edition*. The copyright page lists a first printing of 500 copies.

In the early 1990s, the Archives began offering "VedaBase," a searchable computer program containing digital copies of "The Complete Teachings of His Divine Grace A. C. Bhaktivedanta Swami Prabhupāda." For more information please see fn. 2 and fn. 33.

30 Ranjit das reports that as of 2002, there are over 1000 recordings in the collection.

31 It is possible that at least some of the editors working on the 1979 English edition were aware and had access to transcripts, based on select passages in the 1979 English publication that offer a reasonable match with the archival recordings. Perhaps the editors identified text in recording transcripts whose language they felt worthwhile to include in an editorial manuscript that had been drawn from a translation of the German publication. An apologetic appraisal could take account of a perceived rush to publish, along with an assessment that the manuscript otherwise appeared amenable to print.

In a series of written and personal exchanges that took place in 2019 with the author, Michael Cremo noted that he located the following passage spoken by Prabhupāda in a recording transcript dated April 19, 1973, that closely resembles the published version in a chapter with the same date: "Yes. In my body there are millions of living entities. In my intestines there are so many worms."

Since this section on that topic offers many close matches, it may be possible to identify other areas that contain passages of a similar quality. For example, Chapter 11 dated May 15, 1973 in the English edition, also offers similarities to the language in a recording transcript with the same date.

Cremo felt that editors working on the English edition of *Life Comes from Life* may have had access to an early set of archival transcripts. He noted that that the Bhaktivedanta Archives established in Los Angeles during the late 1970s, had accumulated a sizable collection of transcripts that he was able to periodically access while working for *Back to Godhead* magazine at the time.

32 A pdf containing images of all four documents comparing the first page contents of chapter two from: (1) the 1974 German edition, (2) the 1970s editorial manuscript, (3) the 1979 English edition, and (4) the 1973 recording transcript, is available at following address: https://bit.ly/LCFL_4documents_comparison. Images are also available in the Appendix.

A two-column document offering a side-by-side comparison of the full text of chapter two from the 1979 English edition chapter with the full text of the April 19, 1973 recording transcript that it was based on, is available at the following address: https://bit.ly/LCFL_2text_comparison.

33 Figure 4 is an image from a recording transcript initially published in the Bhaktivedanta Archives' twenty-volume set, *Conversations with Śrīla Prabhupāda*. The morning walk conversation dated April 19, 1973 identified with the second chapter can be found in, "Volume Three: London, August 18, 1971 to Los Angeles, April 21, 1973." See: Prabhupāda (1989), pp. 326–341.

The "Preface" to Volume Three states: "This is one of a series of books published by the Bhaktivedanta Archives, a division of the BBT. This series consists of verbatim transcriptions of all the Morning Walk and Room Conversations in the Archives tape library. They are presented unedited and unabridged to give the reader a sense of what it was like to be personally present with Śrīla Prabhupāda at the time" (p. ix).

34 The websites for the Bhaktivedanta Archives VedaBase (vedabase.com) and Bhaktivedanta Library Services Vanipedia (vanipedia.org) are two excellent examples.

The Bhaktivedanta Archives offers downloads of the VedaBase computer program at the following address: https://www.Prabhupāda.com/Vedabase/VedaBase.html.

35 A complete version of the editorial document can be viewed at: http://surrealist.org/jpgspdfs/lcfl.pdf, access date 25 July 2023. See pages 11–25 in the pdf document for "Morning Walk II" which corresponds with the second published chapter. Both are identified with the morning walk that took place on April 19, 1973.

The pdf copy of the editorial manuscript is located on a website belonging to Nori Muster, the author of the exposé *Betrayal of Spirit* (Urbana: University of Illinois Press, 1997). Muster described the circumstances by which she came to possess these documents in an email exchange with the author:

"After some time I decided to post the scans under the theory that something is better than nothing. That manuscript came to me because it was in the file cabinets at the ISKCON p.r. office in Los Angeles. I worked there for ten years, as described in my book. When our office broke apart in about 1987, everyone went in different directions. I ended holding the bag with all the files somehow. Some paper[s] I returned to people still involved in ISKCON, but the rest I kept. Eventually, I turned it all over to the J. Gordon Melton Collection at the UCSB Special Collections Library."

Muster held the impression that the raw editorial manuscript offered an accurate transcription of the 1973 conversation recordings, as compared to it being an English translation of the 1974 German edition, *Leben kommt von Leben*. She continued: "The LCFL manuscript shows how heavily they edited Śrīla Prabhupāda's dictation tapes. From my experience working with the p.r. department and L.A. BBT, they would make copies of manuscripts and pass them around to an editorial committee. In this case, I would guess they may have given a copy to Ramesvara, Radha-vallabha, Mukunda, and anyone else mentioned in the acknowledgements. The editing marks could have already been there when they made the copies. The copy I got was probably what was given to Mukunda, since the p.r. department was his office, and most of the paper in the files was his." (Muster to Kreitzer, 22 March 2014).

36 The text of the English editorial manuscript sourced to offices affiliated with the Los Angeles BBT, more closely corresponds to the 1974 German publication than to the original recordings. As pointed out in the text – a straightforward obvious example involves the 1974 German publication and the translated editorial draft misspelling "Cheviot Hills Park" as "Shevot-Hills Park," with the latter misspelling a phonetic mistake as might be considered if done by German translators unfamiliar with the Los Angeles area. The Los Angeles BBT editors used the correct spelling in the 1979 English language edition.

37 A. C. Bhaktivedanta Swami Prabhupāda's 1972 *Bhagavad-gītā As It Is: Complete Edition* reads: "The material nature is working under My direction ..." The 1983 *Bhagavad-gītā As It Is: Second Edition* reads: "This material nature, which is one of My energies, is working under my direction ..."

38 The four texts do contain a few short phrases that offer similar matches. They can be located with search terms such as "in the future" and "post-dated check."

39 See fn. 32 for a link to a document offering a side-by-side comparison of the 1979 publication with the transcription for the 1973 audio recording.

40 See the last two pages of the document described in fn. 32.

41 With there being no archival references listed for the date identified with chapter one, its possible that chapter two could well be the first actual morning walk dedicated to the topic. If so, it would offer a natural opportunity for Prabhupāda to explain his objective for entering these discussions.

42 Prabhupāda would identify four basic defects essential to conditional experience as birth, death, disease, and old age.

43 For a link to the entire recording of the morning walk conversation along with a recording transcript, see: Bhaktivedanta VedaBase, "Morning Walk at Cheviot Hills Park, April 19, 1973, Los Angeles," https://vedabase.io/en/library/transcripts/730419mwla/, access date 25 July 2023.

44 For a link to the Vaniquotes.org website offering the specific excerpt under discussion, see: bit.ly/Vaniquotes_LCFL_point, access date 25 July 2023.

Note: the word "vehemently" at the end of the first paragraph of the transcription text, appears clearly spoken as "vibrantly" in the audio recording.

Vertical Dimension and Size of the Purāṇic Universe: Current Interpretations and New Insights

Vasyl Semenov and Evgeniya Semenova
Kiev Krishna Consciousness Academy

Presented at the
Modern Journey Through Ancient Cosmography Seminar
November 21–22, 2020

"Śrīmad-Bhāgavatam *is as great as Kṛṣṇa, the Supreme Lord and shelter of everything. In each and every verse of* Śrīmad-Bhāgavatam *and in each and every syllable, there are various meanings.*"[1]

1. Introduction

In this investigation, we are considering some questions related to the measurements of the vertical dimension of the Purāṇic universe. The work is based on A. C. Bhaktivedanta Swami Prabhupāda's purports on the Fifth Canto of the *Śrīmad-Bhāgavatam,*[2] *Commentaries on the Fifth Canto by Vaiṣṇava Ācāryas*[3] edited by Danavir Goswami, cosmology books by Dr. Richard L. Thompson (Sadāpūta),[4,5] and the work of Bhaktisvarupa Dāmodara Goswami.[6]

The object of investigation in the Fifth Canto is the "Brahmāṇḍa," which is in some way related to the modern notion of the "universe." The *Bhāgavatam* describes innumerable Brahmāṇḍas, each one with its own spherical shell containing an earth disk, called Bhū-maṇḍala. Bhū-maṇḍala divides the universe in half, forming an upper, heavenly region and a subterranean region filled with water. As shown in Figure 1, the Brahmāṇḍa is surrounded by 8 layers of material elements (earth, water, fire, air, ether, mind, intelligence, and ego) with each successive layer being 10 times thicker than the previous one.

The diameter of our Brahmāṇḍa, as given in the Śrīmad-*Bhāgavatam*[7], is 500,000,000 *yojanas* (~6.5 billion km or 4 billion mi). Thus, the radius of Bhū-maṇḍala, the main component of the universe's horizontal structure, is

250,000,000 *yojanas* (~3.25 billion km or 2 billion mi), which is comparable to the distance to Uranus (~2.9 billion km or 1.8 billion mi). Bhū-maṇḍala can be interpreted in many different ways, four of which are found in *Mysteries of the Sacred Universe* (MSU) by Sadāpūta Prabhu.[8]

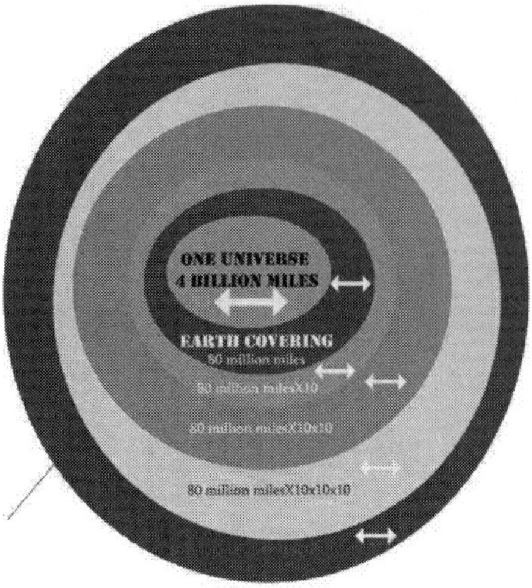

Figure 1. Brahmāṇḍa surrounded by coverings (not to actual scale).

The "vertical" dimension of the Brahmāṇḍa, however, is more difficult to account for than the horizontal dimension. The vertical features of the Purāṇic universe are mainly discussed in Chapter 22–24 of the Fifth Canto and are said to consist of 14 layers spread out over its 500,000,000-*yojana* diameter. These dimensions are confirmed in other *Purāṇas* and in the *Caitanya-caritāmṛta*.[9] This size of the Brahmāṇḍa is relatively small in comparison with the modern estimation of the observable universe[10] (even if we take into account the coverings of the Brahmāṇḍa). The corresponding data are given in Table 1.

Table 1. The sizes of the Brahmāṇḍa, Milky Way, and observable universe.

Object	Diameter in light years
Brahmanda	0.0007 (6 light hours)
Brahmanda + 3 coverings	0.8
Brahmanda + 7 coverings	7620
Brahmanda + 8 coverings	76,208
Milky Way	~100,000
Observable universe as per modern estimation	~93 billion

Even more problematic than the size of the universe are the *heights* of the planets given in *Śrīmad-Bhāgavatam*. As shown in Table 2, these distances differ significantly from modern mean distances to these planets.[11] We used a value of 13.667 km (8.489 mi) for the *yojana,* as suggested by Sadāpūta.[12]

Table 2 Purāṇic heights and mean distances to the planets.

Object	Puranic height (x1000 yojanas)	Modern mean distance (x1000 yojanas)
Sun	100	10951
Moon	200	28
Stars	400	2.94×10^9 (nearest, 4.24 light years)
Venus	600	10959
Mercury	800	10951
Mars	1000	16679
Jupiter	1200	56957
Saturn	1400	104488

What do these heights and the size of the universe mean? In the following section, we consider several existing explanations and then propose two new ones.

2. Previous Work

2.1 Variations on the Length of a Yojana

A commonly assumed value for the *yojana* is approximately 13 km (8–8.5 mi). Nonetheless, we may not understand the distances "as they are" correctly. For example, it is not clear how to adjust the Brahmāṇḍa's diameter of 500,000,000 *yojanas* given in the *Bhāgavatam* with its circumference of 18,712,069,200,000,000 *yojanas* as given by Śrīla Bhaktisiddhānta Sarasvatī, and included by Śrīla Prabhupāda in his purport for *Caitanya-caritāmṛta.*[13] The idea that the latter value is for the total diameter of the universe with its coverings is supported neither by Sadāpūta, nor by the interpretation suggested in the abovementioned purport.

There are different estimations of the size of the *yojana* (e.g., Sadāpūta in *Vedic Cosmography and Astronomy*[14]). A value of ~12.9 km (8 mi) per *yojana*, however, is generally accepted within ISKCON because of Śrīla Prabhupāda's endorsement. Sadāpūta fine-tuned this value to 13.6 km (8.49 mi) per *yojana* because this value produced a striking closeness-of-fit between the modern geocentric orbits of the planets and the rings of Bhū-maṇḍala. Still, the issue with planetary heights suggests that the term *yojana* may imply different values under various circumstances.

2.2 Alternative Understandings of Distance

To resolve these types of discrepancies, some researchers suggest that the distances described in the *Bhāgavatam* are the distances experienced by the yogis, *ṛṣis* and other higher beings. If the *Śrīmad-Bhāgavatam* explains "unearthly" philosophy, it is logical that the Purāṇic structure of the universe is also presented from the viewpoint of entities who have a higher level of consciousness. This point is additionally substantiated by Śrīla Prabhupāda's words "time and space are two correlative terms."[15] If time is perceived in a different way at the different levels of the universe, the same could be true with the perception of distances. Fig. 2 shows how the seven upper worlds and the polar axis (aligned along Mount Meru) are compared with the seven spinal *cakras* and the spinal column of the yogi.

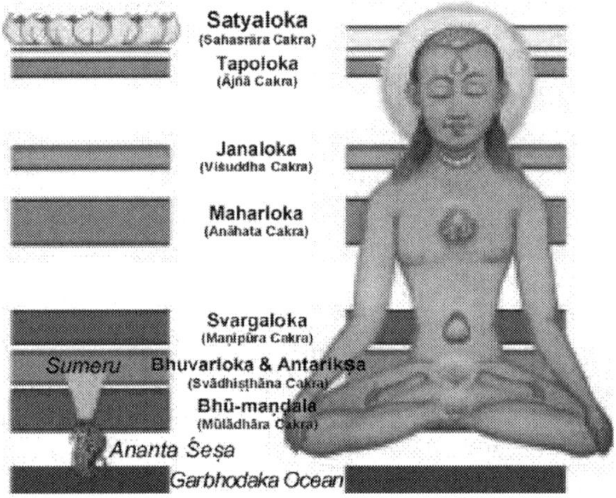

Figure 2. Graphical representation of Purāṇic lokas and yogic cakras.

Similar to the dilation of time from the perspective of the demigods, it is also possible that space itself takes on different proportions in different parts of the universe. The idea that physical laws may be different in various points of time-space is suggested by the general theory of relativity and elaborated upon by Sadāpūta in *Mysteries of the Sacred Universe.*[16]

Another option is that these distances relate to an unknown dimension outside of the purview of traditional 4D space-time. This explanation is often used by Sadāpūta. The multidimensional approach to the universe is currently being developed by string theorists. Additionally, the theory of branes also suggests that some dimensions of the universe may be quite small.[17] Still, nothing is known about the origin or nature of these alternative dimensions. The physical reality that we observe could be the projection of a more voluminous reality, one with many more dimensions, onto our limited senses. Just as a city map is not a city, the universe we see is a map of a different, more voluminous reality, presented as a three-dimensional projection. As Sadāpūta writes, "...the Purāṇic Brahmāṇḍa can be seen as a system of *higher-dimensional* heavenly worlds, combined with an astronomically reasonable model of the solar system."[18]

The *Purāṇas* also describe special "routes" in the universe (e.g. the descent of the Ganges, travels of Durvāsā and Śiva, Suṣumṇā ray, etc.). In the purport by Vijayadhvaja Tīrtha to the *Śrīmad-Bhāgavatam*, Canto 5, Chapter 20, Verse 35,[19] distance is associated with the degrees of happiness. Ashish Dalela (Rsiraja Das), in his 2016 book *Mystic Universe,*[20] takes a similar kind of semantic approach by suggesting at least four different ways to redefine "distance" according to different contexts. These definitions of distance include (a) the linear distance between two objects, (b) the amount of effort it takes to transform an object into another (c) the time it takes to send information from one to another, and (d) the length of the *path* it would take for an observer to *become* that object.

2.3 The Astrolabe Model

The uniform spacing of 200,000 *yojanas* for celestial bodies (Table 1) could be a hint that these are not factual distances, but may simply refer to layers in astrolabe-like models. An exploded diagram of an astrolabe is shown in Figure 3. This is another interpretation suggested by Sadāpūta in *Mysteries of the Sacred Universe*[21]

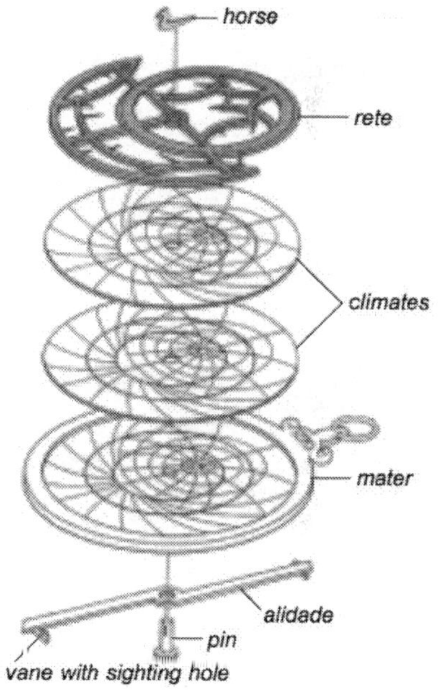

Figure 3. Exploded picture of an astrolabe, showing plates representing the earth (*mater*) and the ecliptic and star positions (*rete*).

2.4 Heights above the ecliptic plane

In both *Vedic Cosmography and Astronomy*[22] and *Mysteries of the Sacred Universe*,[23] Sadāpūta Prabhu interpreted the heights of the planets to be the maximum distance a planet assumes above and below the ecliptic plane due to its orbital inclination, shifted by 100,000 *yojanas*. A schematic diagram of this concept is shown in Figure 4. The following transformation formula can be applied to express this value in *yojanas*:

$$H = \frac{h}{8.489 \times 1.61} + 100000$$

where H is the height above Bhū-maṇḍala in *yojanas*, and h is the maximum distance of a planet from the ecliptic plane in kilometers, using the conversion factors of 8.489 miles per *yojanas* and 1.61 kilometers per mile. The corresponding data are presented in Table 3.

Figure 4. Interpretation of planetary heights as given in MSU.

Table 3. Purāṇic heights along with their estimations, based on the heights of the planets above the ecliptic.

Planet	Puranic height (H_p, in thousands of yojanas)	Height above ecliptic (*h*, in millions of kilometers)	Transformed height (*H*, in thousands of yojanas)
Sun	100	0	100
Moon	200	0.03	102
Venus	600	6.34	564
Mercury	800	6.33	563
Mars	1000	8.09	692

Planet	Puranic height (H_p, in thousands of yojanas)	Height above ecliptic (*h*, in millions of kilometers)	Transformed height (*H*, in thousands of yojanas)
Jupiter	1200	19.72	1543
Saturn	1400	65.70	4907
Correlation		0.76	0.76

As shown in Table 3, there is a good correlation between Purāṇic heights (column 2) and transformed heights above the ecliptic (column 4), except for the moon and Saturn. For a deeper statistical analysis, however, we propose using correlation coefficients (also known as the Pearson correlation coefficient[10]), which is defined for two vectors, $x = (x_1, x_2, x_3, \dots x_n)$ and $y = (y_1, y_2, y_3, \dots y_n)$ as follows:

$$K_c(\mathbf{x},\mathbf{y}) = \frac{\sum_{k=1}^{n}(x_k - \overline{x})(y_k - \overline{y})}{\sqrt{\sum_{k=1}^{n}(x_k - \overline{x})^2}\sqrt{\sum_{k=1}^{n}(y_k - \overline{y})^2}}, \qquad (1)$$

where

$$\overline{x} = \frac{1}{n}\sum_{k=1}^{n} x_k ,$$

and

$$\overline{y} = \frac{1}{n}\sum_{k=1}^{n} y_k$$

are mean (average) values of vectors **x** and **y** respectively.

The values of a correlation coefficient lie in the interval [0,1]. The value of the correlation coefficient represents the probability that the variance observed in the dataset is not random.

The values for the correlation coefficients $K_c(H_p,h)$ (between Purāṇic heights and heights above the ecliptic) and $K_c(H_p, H)$ (between Purāṇic heights and transformed heights) are both equal to 0.76 (the last row of table

3). Note that $K_c(H_p,h) = K_c(H_p,H)$ because the correlation coefficient is not affected by a linear vector transformation. The correlation coefficient being equal to 0.76 reveals a *moderate* linear dependence between Purāṇic heights and heights above the ecliptic, though the small amount of data (i.e., just seven elements) does not allow us to make definitive conclusions.

If we graph Purāṇic heights vs heights above the ecliptic on a coordinate plane (Fig. 5), we see that they are quite scattered in relation to the line which shows the *optimal linear prediction* of Purāṇic heights based on heights above the ecliptic. This confirms that Purāṇic heights and heights above the ecliptic have only a slight linear dependence.

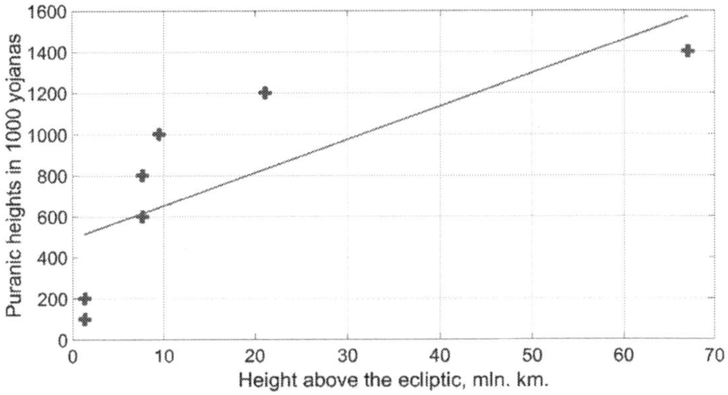

Figure 5. Graphical representation of Purāṇic heights and heights above the ecliptic.

3. New Perspectives

3.1 Correlation with Modern Distances to the Planets

As explained by Sadāpūta, from the point of view of someone on earth, the planets orbit in epicycles (see for example, the geocentric orbit Mercury shown in Fig. 6). Thus, for the geocentric orbit of each planet we can find the following values:

minimum distance from the earth, d_{min}

maximum distance from the earth, d_{max}

mean distance from the earth, $0.5(d_{min} + d_{max})$

difference between maximum and minimum distances, $(d_{max} - d_{min})$

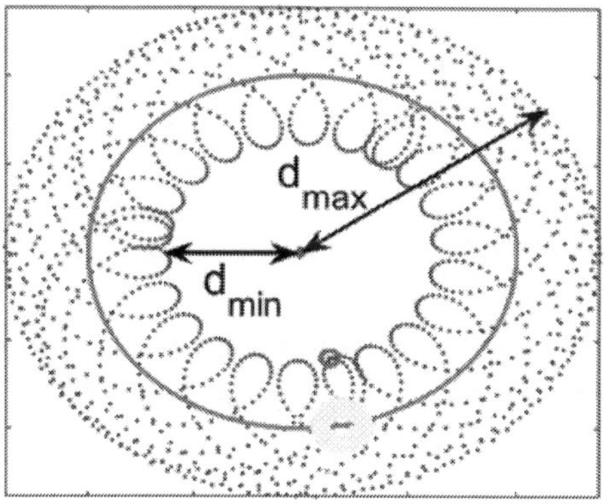

Figure 6. Example of the epicycle trajectory of Mercury around the earth (dot in the center). The sun is represented by the yellow disk.

Table 4 presents the data for mean, minimum, and maximum distances to the planets along with the corresponding correlations with Purāṇic heights. The correlation values for these distances (0.73, 0.84, 0.80) are comparable to the correlation for the heights above the ecliptic in Table 3 (0.76). Yet, if we consider the correlation of Purāṇic heights with the *difference between maximum and minimum distances to the planets* (column 5), the value of the correlation coefficient is much higher, namely, 0.96, which implies a strong linear dependence between Purāṇic heights and the values for (d_{max}- d_{min}).

This dependence between Purāṇic heights and "max-min differences" is illustrated in Fig. 7. The line shows the optimal linear prediction of Purāṇic

Table 4. The parameters of planetary orbits (in 1000 *yojanas*) as given in MSU,[24] along with their respective correlation coefficients.

Planet	Minimum distance (d_{min})	Maximum distance (d_{max})	Mean distance (d_{mean})	Difference of max-min distances
Sun	10761	11140	10951	379
Moon	26.9	29.6	28	3
Venus	2903	19014	10959	16111
Mercury	6019	15883	10951	9864
Mars	4075	29282	16679	25207
Jupiter	43315	70599	56957	27284
Saturn	87237	121739	104488	34502
Correlation	0.73	0.84	0.80	0.96

heights based on "max-min differences." It is interesting to note that this line passes almost exactly through the points corresponding to the moon and Saturn. The corresponding linear predictions for Purāṇic heights based on "max-min differences" are given in Table 5 (column 4), along with Purāṇic heights (column 2) and approximations based on heights above the ecliptic (column 3) for the purpose of comparison.

Besides the correlation coefficients that were given earlier in Tables 3 and 4, we also provide the mean error, which is defined for two vectors $x = (x_1, x_2, x_3, \dots x_n)$, $y = (y_1, y_2, y_3, \dots y_n)$, as follows:

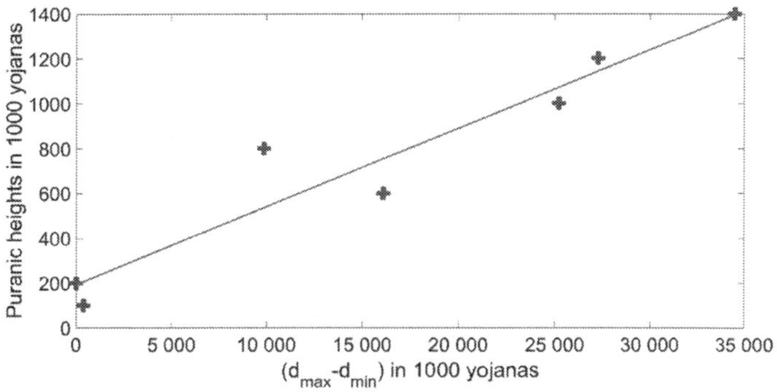

Figure 7. Graphical representation of Purāṇic heights and "max-min differences.

Table 5. Comparison of Purāṇic heights with their approximations based on heights above the ecliptic and "max-min differences" (in 1000 *yojanas*).

Planet	Puranic height (H_p)	Transformed height (*H*)	Transformed "max-min difference"
Sun	100	100	206
Moon	200	102	193
Venus	600	564	754
Mercury	800	563	537
Mars	1000	692	1071
Jupiter	1200	1543	1143
Saturn	1400	4907	1395
Correlation		0.76	0.96
Error		647.0	94.8
Normalized error		0.56	0.25

$$E(\mathbf{x},\mathbf{y}) = \frac{1}{n}\sum_{k=1}^{n} | x_k - y_k |$$

and average error

$$E_{norm}(\mathbf{x},\mathbf{y}) = \frac{1}{n}\sum_{k=1}^{n} \frac{| x_k - y_k |}{x_k}$$

Note that the normalized error is similar to the one used by Sadāpūta in MSU for the correlation of geocentric planetary orbits with Bhū-maṇḍala features.[25]

Table 5 shows that the linear approximation based on transformed "max-min differences" not only provides a higher correlation coefficient but also provides lower values of error and average error (6.8 and 2.2 times, respectively) compared to the approximation based on the heights above the ecliptic.

In this subsection, we have shown the connection of Purāṇic heights to the differences between the maximum and minimum distances to the planets in a geocentric orbit. The approximation of Purāṇic heights by this approach was shown to be more exact than the comparison based on the heights above the ecliptic, both in terms of correlation coefficients and mean errors. The values for the planetary heights given in the *Bhāgavatam,* however, have the same order of magnitude as their respective maximum distances from the ecliptic plane. This is not the case when correlating the heights to the distance from the plane of the ecliptic. But that approach does not explain such issues as the Purāṇic height of the moon and the size of the Brahmāṇḍa.

3.2 Correlation of Purāṇic heights with log-distances

We also calculated the correlation coefficients between Purāṇic planetary heights and the logarithms of mean distances to the planets as given in Table 4, column 4. The corresponding logarithms of mean distances are given in Table 6, column 3. The log-distances along with Purāṇic heights are represented in Fig. 8. Although we see from Fig. 8 that the data are quite scattered, which is confirmed by the relatively small correlation coefficient (0.71), it can also be seen that points for the last four planets (i.e., starting from Mercury) lie approximately on the same line (with correlation coefficient 0.98). The line in Fig. 8 shows the optimal linear prediction of Purāṇic heights based on the logarithmic mean distances for the last four planets. The equation of this line is:

$$y = p_1 \log(d_{mean}) + p_2, \qquad (2)$$

where $p_1 = 241780$ and $p_2 = -3712600$.

Table 6. Purāṇic heights along with logarithms of mean distances to the planets.

	Puranic height (in 1000 yojanas)	Log-mean distance (in kilometers)
Sun	100	18.8
Moon	200	12.9
Venus	600	18.8
Mercury	800	18.8
Mars	1000	19.2
Jupiter	1200	20.5
Saturn	1400	21.1
Correlation		0.71
Correlation for the last 4 planets		0.98

Since astronomical distances are often expressed in light years, let us introduce the corresponding value:

$$d_{mean}^{LY} = \frac{d_{mean}}{c \times 365 \times 24 \times 3600},$$

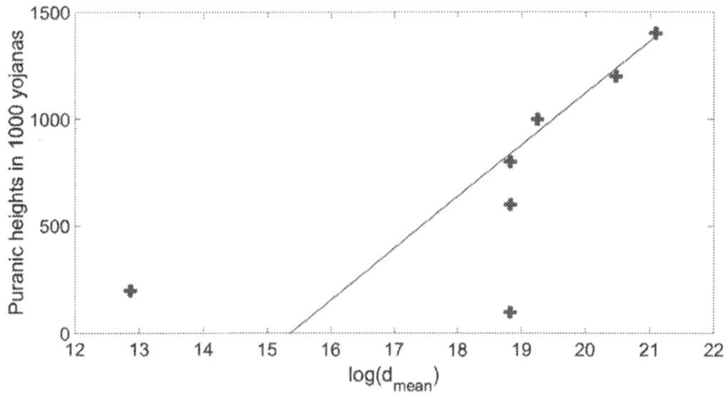

Figure 8. Graphical representation of Purāṇic heights and logarithmic mean distances to the planets.

Where $c = 3\times10^5$ km/s is the speed of light.

Hence, formula (2) can be rewritten as

$$y = p_1 \log(d_{mean}^{LY}) + p_2', \qquad (3)$$

where $p'_2 = 3511213$

We can also use this approximate linear dependence to estimate, for example, the Purāṇic height of Dhruvaloka (the Pole Star, currently Polaris), the distance to which is estimated today as 433 light years. According to formula (3), the predicted value is 4,980,000 *yojanas*, which has the same order of magnitude as the value of 3,900,000 *yojanas*, the height given for Dhruvaloka.[26] Similarly, if we the take the radius of the *observable universe* estimated as 93 $\times 10^9$ light years,[10] we obtain an approximate estimate of 9,600,000 *yojanas*, which has the same order of magnitude as the height of Maharloka (about 30% less), given as 13,900,000 *yojanas*.[26]

4. Conclusion

In this essay we have unpacked some possible meanings for the concept of "heights above Bhū-maṇḍala" as described in the 5th Canto of the *Śrīmad-Bhāgavatam*, to see how they may relate to the world we experience. First, we examined various ways this has been understood within ISKCON. Those who understand the distances "as they are" encounter inexplicable conflicts with simple observations and generally invoke *acintya-bhedābheda-tattva* to justify this position. Taking a more indirect approach, we explored several interesting analyses given in terms of higher dimensionality, space dilation, and levels of consciousness. We touched upon the astrolabe model suggested by Richard L. Thompson in *Mysteries of the Sacred Universe,* and examined in more depth the correlation between planetary heights and orbital inclinations that he investigates first in *Vedic Astronomy and Cosmography* and then develops further in *Mysteries of the Sacred Universe.* Asish Dalela, in his book *Mystic Universe,* proposes several novel ways to understand concepts such as "distance" from a Vedic viewpoint that help to broaden our understanding.

Finally, noting that the *yojana* assumes different values under various circumstances, we researched correlations between Purāṇic planetary heights and distances from Earth as currently understood by modern astronomy. By using Pearson correlation coefficients, we showed a strong linear dependence between Purāṇic heights and the values for the max-min differences of the planetary orbits around the earth. The correlation was significantly greater than the linear correlation with orbital inclinations above the ecliptic. We also show a correlation with the logarithms of mean distances to the planets.

Although in this essay we establish significant parallels between the *Śrīmad-Bhāgavatam's* description of the Brahmāṇḍa and the universe of our experience, there is clearly no simple one-to-one mapping between the two. This is to be expected, for as Śukadeva Gosvāmī explains to Mahārāja Parīkṣit as he begins his description of Jambūdvīpa:

> *My dear King, there is no limit to the expansion of the Supreme Personality of Godhead's material energy. This material world is a transformation of the material qualities* [sattva-guṇa, rajo-guṇa, *and* tamo-guṇa], *yet no one could possibly explain it perfectly, even in a lifetime as long as that of Brahmā. No one in the material world is perfect, and an imperfect person could not describe this material universe accurately, even after continued speculation. O King, I shall nevertheless try to explain to you the principal regions, such as Bhūloka, with their names, forms, measurements and various symptoms.*[27]

References

1 A. C. Bhaktivedanta Swami Prabhupada, *Sri Caitanya-caritāmṛta, Madhya-līlā*, Chapter 24, Verse 318 (Los Angeles, CA: The Bhaktivedanta Book Trust, 1975), p. 260.

2 A. C. Bhaktivedanta Swami Prabhupada, *Śrīmad-Bhāgavatam*, Canto 5 (Los Angeles, CA: The Bhaktivedanta Book Trust, 1988).

3 *Commentaries on 5th Canto Bhāgavatam Cosmology*, edited by Danavir Goswami (Kansas City, MO: Rupanuga Vedic College, 2007).

4 Richard L. Thompson, *Vedic Cosmography and Astronomy* (Los Angeles: Bhaktivedanta Book Trust, 1989).

5 Richard L. Thompson, *Mysteries of the Sacred Universe: The Cosmology of the Bhāgavata Purāṇa* (Alachua, FL: Govardhan Hill Publishing, 2000).

6 Bhaktisvarūpa Dāmodara Goswami, *Life and Origin of the Universe* (Kolkata, India: Bhaktivedanta Institute, 2004).

7 Prabhupāda, *Śrīmad-Bhāgavatam*, Canto 3, Chapter 11, Verses 40–41 (Los Angeles, CA: The Bhaktivedanta Book Trust, 1988), pp. 490–91.

8 Thompson, *Mysteries of the Sacred Universe* (Alachua, FL: Govardhan Hill Publishing, 2000), p. 3.

9 Prabhupāda, *Sri Caitanya-caritāmṛta, Madhya-līlā*, Chapter 21, Verse 84 (Los Angeles, CA: The Bhaktivedanta Book Trust, 1975), p. 276.

10 I. Bars; J. Terning, F. Nekoogar, *Extra Dimensions in Space and Time* (New York, NY: Springer, 2010), p. 27.

11 Thompson, *Mysteries of the Sacred Universe* (Alachua, FL: Govardhan Hill Publishing, 2000), p. 121.

12 Ibid., Chapter 4, p. 99.

13 Prabhupāda, *Sri Caitanya-caritāmṛta, Madhya-līlā*, Chapter 21, Verse 84 (Los Angeles, CA: The Bhaktivedanta Book Trust, 1975), p. 276.

14 Thompson, *Vedic Cosmography and Astronomy* (Los Angeles: Bhaktivedanta Book Trust, 1989), pp. 24–25.
15 Prabhupāda, *Śrīmad-Bhāgavatam*, Canto 3, Chapter 11, Verse 4 (Los Angeles, CA: The Bhaktivedanta Book Trust, 1988), p. 461.
16 Thompson, *Mysteries of the Sacred Universe* (Alachua, FL: Govardhan Hill Publishing, 2000,) Chapter 4, pp. 121–22.
17 L. Randall, *Warped Passages* (New York, NY: Harper Perennial, 2006), p. 37.
18 Thompson, *Mysteries of the Sacred Universe* (Alachua, FL: Govardhan Hill Publishing, 2000), Chapter 7, p. 198.
19 *Commentaries on 5th Canto Bhāgavatam Cosmology*, edited by Danavir Goswami (Kansas City, MO: Rupanuga Vedic College, 2007), p. 433.
20 Ashish Dalela, *Mystic Universe: An Introduction to Vedic Cosmology* (Duarte, CA: Shabda Press, 2016).
21 Thompson, *Mysteries of the Sacred Universe* (Alachua, FL: Govardhan Hill Publishing, 2000,) Chapter 3, pp. 59–60, 67.
22 Thompson, *Vedic Cosmography and Astronomy* (Los Angeles: Bhaktivedanta Book Trust, 1989), Chapter 4, p. 85.
23 Thompson, *Mysteries of the Sacred Universe* (Alachua, FL: Govardhan Hill Publishing, 2000), Chapter 4, pp. 120–21, Chapter 6, pp. 175–76.
24 Ibid., Chapter 4, p. 98.
25 Ibid., Appendix 7, pp. 263–67.
26 Prabhupada, *Śrīmad-Bhāgavatam*, Canto 5, Chapter 23, Text 9 (Los Angeles, CA: The Bhaktivedanta Book Trust, 1988), p. 832.
27 Ibid., Chapter 16, Text 4 , p. 557.

The Correlation Between Materials Science and the Vedic System of Sounds

Fedor Vodolazskiy
Ural Federal University

Presented at the
Consciousness in Science Conference
January 18–20, 2019

Abstract: The idea of confirming the cosmology and the structure of the world described in the *Śrīmad-Bhāgavatam* and other *śāstra* by modern science was proposed by Śrīla Prabhupāda. The idea of this paper is to find the correlation between materials science and the Vedic description of reality through a system of sounds. There is mathematical theory modeling the theory of phase transformations in metal-based alloys by using four-dimensional space. The example is crystallography. Since the 1980's, special geometric constructions have been used to describe the atomic structure of condensed phases. These constructions were four-dimensional polyhedrons, called polytopes. In this case, the four-dimensional structure determined the positions of atoms of the three-dimensional lattice. The four-dimensional structure in this case is an analogue of the sound level *madhyama* which determines the three-dimensional structure, the level of sound *vaikharī*. Thus, such a description allows us to find the correlation between materials science and the Vedic system of sounds. The search for such analogies in materials science can improve the authority of the Vedic description of reality, based on the *Śrīmad-Bhāgavatam* and other *śāstra.*

The idea of confirming the cosmology and the structure of the world described in the *Śrīmad-Bhāgavatam* and other *śāstra* using modern science was proposed by Śrīla Prabhupāda. The idea of this paper is to find the correlation between materials science and the Vedic description of reality through a system of sounds. It is well known that Vedic system of sounds consist of four levels:

para
paśyanti
madhyama
vaikharī

According to the *Ṛg Veda* (1.164.45), the first three levels remain hidden within the heart as imperceptible vibrations, which indicates that they are not perceived by our senses. The *vaikharī* level of sound is the ordinary sound we can hear. If we accept that the matter around us represents a *vaikharī* level of sound, then modern science indirectly corroborates this, since in quantum mechanics it is assumed that only vibrations exist at the subatomic level.[1, 2] So, how do we understand and describe the first three levels of sound? These levels are inaccessible to our senses, but to some extent they can be described with the help of logic. Our senses can perceive three-dimensional objects, but four and more multidimensional objects cannot be fully perceived – our senses can see only three-dimensional projections of multidimensional objects. This vision of the world is well-illustrated in the book *Flatland* by Edwin Abbott (1884), which describes how the three-dimensional objects of our world look from the point of view of a two-dimensional being. Similarly, we can extrapolate how the objects in the three-dimensional world may be part of more complex multidimensional objects that are not fully perceived by the senses. So we will use scientific evidence of multidimensionality in our universe as analogous to the sound levels above *vaikharī*. The description of multidimensional objects is quite easily done with the help of mathematics by introducing another dimension using complex numbers. This paper presents an attempt to present how to describe three-dimensional objects in materials science, and specifically in crystallography, using sections of multidimensional objects.

The traditional crystallography approach is to use 14 types of Bravais lattices and describe the atomic position with these types.[3] The most typical examples in metal are cubic and hexagonal lattices. Ordinarily they have a 2, 3, 4, or 6-fold axis of symmetry. The 2-fold axis of symmetry means that

1 Richard P. Feynman, Robert B. Leighton, and Matthew Sands, *The Feynman Lectures on Physics*, Vol. III: The New Millennium Edition: Quantum Mechanics, p. 400

2 Becker, Katrin; Becker, Melanie; Schwarz, John; *String Theory and M-theory: A Modern Introduction* (2007), Cambridge University Press, p. 757.

3 Kraposhin V.S. and Talis A.L., *Crystallography and Matter*, Priroda (2014), №11, p. 3–15 (in Russian)

atoms move to the same position after a 180° rotation. For the 3-fold axis of symmetry, the angle is 120°, and so on. A cubic lattice has a 2, 3, or 4-fold axis of symmetry, and the hexagonal lattice has a 2, 3, or 6-fold axis of symmetry. These unit cells are units of structure that can fill the space without gaps using translation of lattices like flats in a high-rise building. But this approach has some difficulties in describing lattices with a 5-fold axis of symmetry. There are many examples of materials in nature whose lattices have a 5-fold axis of symmetry. These materials are widely known as quasicrystals. The problem is the 5-fold axis of symmetry is compatible with translation, but only in one direction – in other directions there is non-periodic symmetry. To describe this structure, we must use another approach.

Since the 1980's, special geometric constructions have been used to describe the atomic structure of condensed phases. These constructions are multidimensional polyhedrons, called polytopes. These polytopes describe atomic position in three-dimensional space. The need for the introduction of four-dimensional space is caused by the following: Let us consider pentagonal tiling on a plane. A regular pentagonal tiling on the Euclidean plane is impossible without gaps (Figure 1a). The solution is to expand into three-dimensional space and fill the gaps with a three-dimensional dodecahedron (Figure 1b). Similarly, filling space with figures for materials with the 5-fold axis of symmetry is possible by using a mathematical expansion into 4-dimensional space. In this case, three-dimensional space is filled without gaps by three-dimensional sections of a four-dimensional polytope with a fifth axis of symmetry. The choice of section is carried out by rotating the four-dimensional polytope. Examples of polytops are illustrated in Figure 2. This figure describes atomic position in three-dimensional space. Atoms are situated on the surface of these figures.

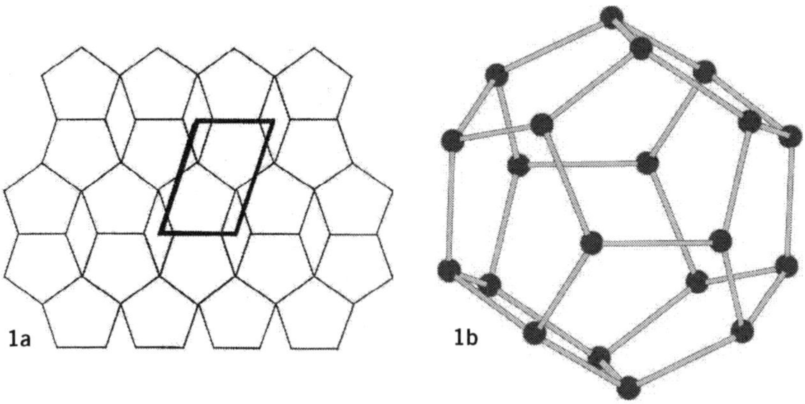

Figure 1. Regular pentagonal tiling on the two-dimensional space (a) and three-dimensional space (b)

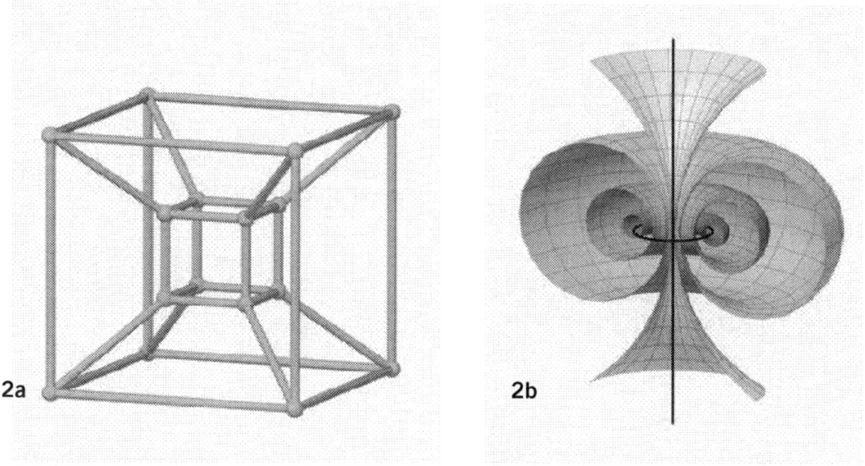

Figure 2. Examples of four-dimensional polytopes: 4D-cube (a); 4D-sphere (b)

Moreover, in 1978 Japanese scientists at the University of Aizu proposed the idea of a prototype phase or prophase.[4] The prophase is the real or hypothetical high symmetrical structure with its substructures corresponding in three-dimensional space to structures of real phases participating in the transformation. The idea is that there is a multidimensional object that contains all possible lattice variants for the particular material. Another fact is the existence of quasicrystals. It is known that quasicrystals are three-dimensional projections of six-dimensional objects.[5] According to this approach, the multidimensional structure determines the positions of atoms in the three-dimensional lattice.

These examples show the possibility of explaining phenomena in nature by describing them as multidimensional objects. Crystallography is shown here as a particular case. Perhaps in the future there will be an opportunity to find a similar way to describe puzzling observations in other areas of science. This approach may also help to clarify and deepen our understanding of how reality is described in the *Vedas* using a system of sounds.

4 K. Aizu, "The Concepts 'Prototype' and 'Prototypic phase' – Their Difference and Others." *Journal of the Physical Society of Japan*, Vol. 44 (1978), p. 683–696.

5 S. V. Vonsovsky and M. I. Katsnelson, *Quantum Solid State Physics*, Берлин: Springer (1989), p. 505.

Notes on Contributors

Michael Cremo (Drutakarma) has been involved with the Bhaktivedanta Institute since 1984, and more recently with the Temple of the Vedic Planetarium. He is co-author, along with Richard L. Thompson (Sadaputa Dasa), of *Forbidden Archeology* (1993), and has authored numerous books since then, including *My Science, My Religion: Academic Papers* (1994-2009). He has given talks at the Royal Institution in London, the Russian Academy of Science in Moscow, and the Indian Institute of Science in Bangalore.

Bhupal Dev has a PhD in particle physics from the University of Maryland and performed postdoctoral work in the UK and Germany before joining the faculty of Washington University. His research specializes in going beyond the standard model of particle physics and cosmology. He has published over one hundred peer-reviewed papers in international journals on these and related topics. He is currently working on the anomalies of the Cosmic Microwave Background and the implications for the evolution of the universe.

Jonathan Edelmann (Janakirama) has a BA in Philosophy from the University of California and a PhD (Religious Studies and Theology) from Oxford University. His book *Hindu Theology and Biology: The Bhagavata Purana and Contemporary Theory* (Oxford University Press, 2012) examines the possibility of dialogue between Hinduism and evolution. Edelmann has written about scientific studies of reincarnation, the theory of karma and *samsara*, the doctrine of *bhakti*, and the *Bhāgavata Purāṇa* commentaries of Jīva Gosvāmin, Śrīdhara Svāmin, and Vaṃśīdhara. He is an Assistant Professor in the Religion Department at the University of Florida.

Christopher Hayton (Krsna-bhajana) holds a BSc in Zoology from Cardiff University, and a MA and PhD in Social Work from Florida State University. He presently serves as a research associate for the Bhaktivedanta Institute for Higher Studies, as well as a production manager for GN Press. He has taught at Florida State University, and has also worked as a clinical social worker in community mental health, and in the field of archeology.

S. E. Kreitzer (Sthita-dhi-muni) assisted Richard L. Thompson (Sadaputa), a founding member of the Bhaktivedanta Institute, in the 1990s. While pursuing graduate studies at the University of Florida in History of Science (PhD 2013), he established the Richard L. Thompson Archives with Bob Cohen and others to facilitate access to Thompson's work. He teaches American History and History of Science at Santa Fe College and the University of Florida.

Howard Resnick (Hridayananda Das Goswami) received his PhD in Sanskrit & Indian Studies from Harvard University in 1996, with his dissertation published as part of the Harvard Oriental Series. He was formally initiated into the Gauḍīya Vaiṣṇava tradition in 1970, and over the next several years opened *bhakti-yoga* centers in over a dozen countries. In the 1980s he helped complete Bhaktivedanta Swami Prabhupāda's translation and commentary on the *Bhāgavata Purāṇa*. In 2013, he established Krishna West to help facilitate ISKCON's outreach to Western audiences.

Evgeniya Semenova obtained her PhD degree in Mathematics from the Institute of Mathematics at NAS in 2009, where she has been working at the Computational Mathematics Department since 2004. She received scholarships from the Institutes of Mathematics at Linz (Austria) and Lubeck (Germany) and took part in the EU AMMODIT program. Since 2016, she has been teaching the ISKCON Disciple Course at the Kiev Krishna Consciousness Academy.

Vasyl Semenov (Dvija Govinda) obtained his PhD degree in Acoustics from the Institute of Hydromechanics at the Ukrainian National Academy of Sciences (2004) and a second PhD in Computational Mathematics from the Institute of Cybernetics at NAS in 2020. Since 1999, he has been working at the Scientific & Production Enterprise "Delta SPE" in the area of telecommunications. He is currently a visiting professor at the American University Kyiv and has been teaching Vedic mathematics and other courses in Kiev Krishna Consciousness Academy since 2013.

Richard L. Thompson (Sadaputa, 1947–2008) was a mathematician, scientist, philosopher, researcher of ancient cosmology, author, and devoted practitioner of *bhakti-yoga*. Thompson received his PhD from Cornell University (1974), where he specialized in probability theory and statistical mechanics. During this time he found inspiration in the philosophy of *Bhagavad-gītā* and became a dedicated follower of Gauḍīya Vaiṣṇavism. Thompson subsequently served as a founding charter member of the Bhaktivedanta Institute, established in 1976.

Alexey Timoschuk (Abhinanda) received an MA in French and English (1996) and a PhD in Philosophy (1999) at the Vladimir Institute of Law, and is currently a Professor of the Humanities Dept., Vladimir Institute of Law. He has published numerous articles and presented at international conferences on the topic of the Vaiṣṇava tradition. He is also a founding member of the Bhaktivedanta Institute in Russia.

Tiziano Valentinuzzi (Giridhari) earned his PhD in Astronomy (2007) at the University of Padua. He has studied galaxy evolution and black matter halos, with a special focus on superdense massive galaxies in the near universe, along with dark matter density profiles in near and deep space. For nearly thirty years, he has studied the Vedic tradition with a special focus on *bhakti-yoga*, *Vāstu Śāstra*, and Vedic science. His research interests at the Bhaktivedanta Institute include cosmology and astronomy in ancient Vedic texts.

Fedor Vodolazskiy earned a PhD in Engineering Sciences at Ural Federal University (UrFU) in 2010, with research on the formation of structure, texture, and properties during rolling of high-alloyed titanium alloys. He has been working with the Bhaktivedanta Institute in Russia since 2017, and is currently an Associate Professor at the UrFU Department of Heat treatment and physics of metals.

Made in the USA
Monee, IL
02 July 2026

56326102R00092